Metropolitan
Microwave Network
Design and Implementation

Michael Wang

Prentice Hall, Englewood Cliffs, New Jersey 07632

Library of Congress Cataloging in Publication Data

WANG, M. (MICHAEL)
  Metropolitan microwave network.

  Includes index.
  1. Digital communications--Design and construction.
  2. Microwave communication systems--Design and
  construction. I. Title.
  TK5103.7.W36 1990   621.38'0413   88-36417
  ISBN 0-13-579723-3

Editorial/production supervision
and interior design: BARBARA MARTTINE
Cover design: 20/20 SERVICES, INC.
Manufacturing buyer: MARY ANN GLORIANDE

The publisher offers discounts on this book when ordered
in bulk quantities. For more information, write:
    Special Sales/College Marketing
    Prentice Hall
    College Technical and Reference Division
    Englewood Cliffs, NJ 07632

Printed in the United States of America

10 9 8 7 6 5 4 3 2 1

ISBN   0-13-579723-3

PRENTICE-HALL INTERNATIONAL (UK) LIMITED, London
PRENTICE-HALL OF AUSTRALIA PTY. LIMITED, Sydney
PRENTICE-HALL CANADA INC., Toronto
PRENTICE-HALL HISPANOAMERICANA, S.A., Mexico
PRENTICE-HALL OF INDIA PRIVATE LIMITED, New Delhi
PRENTICE-HALL OF JAPAN, INC., Tokyo
SIMON & SCHUSTER ASIA PTE. LTD., Singapore
EDITORA PRENTICE-HALL DO BRASIL, LTDA., Rio de Janeiro

CONTENTS

CHAPTER 1
THE METROPOLITAN NETWORK

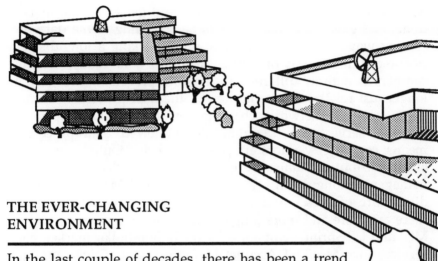

THE EVER-CHANGING
ENVIRONMENT

In the last couple of decades, there has been a trend
toward decentralization in information gathering , processing, and dissemination. By this, I don't mean that organizations have been moving toward a decentralized structure
or that the decision process has been passed on to the lower levels.
What is observed is that organizations are being made up of very
specialized groups, each with their unique objective and expertise.
They gather the information that is unique to them, prepared for
their own use. Very often this information is messaged and distributed to the rest of the organization. As a result, the different raw
data are stored over multiple databases, and the movement of this
information becomes very critical to an organization.

The computer used to be the center of an information system, while telecommunications was the engineering activity that
brought data into and out of the computer. In the last few years,
the balance shifted. Perhaps it was when the installed value of
workstations and personal computers superseded that of main-

frames, perhaps it was when the ratio of computer users with their own keyboards to computer professionals passed the 50-to-1 mark, or perhaps it was when the amount of installed computer processing power devoted to transaction processing passed 50 percent. At any rate, the center of an information system is evolving from a computer to a network. The resource we need to take care of is no longer the machine that adds, multiplies, and divides, but the data it serves.

Remember the time when the data processing department was the only place where anybody could get any data analysis done, and everything had to be done according to the data processing way? Remember the time when making a long-distance phone call meant only one thing, using Ma Bell. Today, the environment is entirely different. In the data processing area, we no longer just rely on one computer, or the data processing department; local area networks chaining personal computers, printers, and file servers are becoming the preferred data processing means; computer vendors are pushing for distributed processing, even distributed databases. In telephony, a telecommunication manager no longer can just rely on good old Ma Bell. With the divestiture of AT&T and the telephone industry deregulations, there are more and more long-distance common carriers, such as MCI and US Sprint, competing for the billion-dollar market. New services such as AT&T Accunet 1.5 high-capacity (high-cap) facility for the first time offer the customers a true digital, high-speed data-voice transport. Then there are talks about a private network, Integrated Service Digital Network (ISDN), along with more and more new technology and new standards emerging.

With the communication environment changing ever so rapidly, how does an organization handle it? The responsibility of the data and voice managers used to be fairly straightforward: to improve a network's overall quality and productivity, lower the recurring cost of the service, and put it under stringent financial control. Unfortunately, the ever-changing communication environment, especially in areas of hardware development, new services, and government regulatory actions, has made this management task very difficult. What a company needs is neither a data processing manager nor a telecommunication manager; instead, it needs an information manager—someone who knows both the data and voice arenas and who knows about private data network, as well as microwave and fiber optics transports—someone who can follow all the development in the industries and give the company an edge in the information management area.

There is an ever-growing recognition by companies on the importance of information networking. We define information networking as telecommunications highways that are used by businesses to collect, analyze, and distribute the information necessary for running the company. The key concept critical to understanding this trend is the time value of information (TVOI) and how TVOI is increasing the desire to communicate more rapidly. The overall drive to communicate faster is being supported technically by the digital network evolution, the availability of higher-speed transmission, and the increasing intelligence in these networks.

The Importance of Information Networks

Companies are becoming more reliant on their ability to gain access to information, and their information networks are being used as strategic weapons as the differentiation among competitors. In response, information networks are moving out of the administrative overhead closet and into the corporate spotlight. Even in this more visible environment with its higher risks, the consolidation of diverse information networks onto larger backbone networks is taking place at an increasing rate because economics makes complex network mapping cost effective.

Telecommunications resources (equipment and facilities) along with data processing resources are increasingly being viewed as strategic corporate assets. For example, Fidelity Investment Corp. in Boston, one of the largest mutual fund companies, instituted an hourly pricing system enabling customers to learn how their mutual funds performed in less than 30 minutes after each hour of a trading day. Customers therefore have more frequent, as well as more current, information with which to make buy and sell decisions, potentially generating more revenue for Fidelity. Federal Express, a giant in the overnight package delivery service business, differentiates itself by providing customers with information on the precise location of their packages in less than 30 minutes. This is done through a massive integration of the communications technology and the data processing technology.

The Time Value of Information

The time value of information, or the rate at which information decays, refers to the degradation in the value of information that occurs over a given period of time. In the broadest sense, information is rapidly developing into a perishable commodity with a decaying halflife, with the rate of decay depending on the specific purpose that the information serves. An example of information with a halflife of almost zero is reflected in the behavior of stocks: the announcement of a company's earnings has value only in the first few minutes that it circulates; however, once the information is widely disseminated, it loses almost all its potency relative to stock price behavior. On the opposite side, information with an almost infinite halflife is Einstein's theory of relativity, which retains value as an underpinning for further theoretical developments in physics.

The TVOI varies according to specific industry dynamics as well as to particular developments within industries. Although there is no reliable method for quantifying the rate at which the value of information is degraded over time, it is intuitively obvious that the information halflife in the investment world is generally shorter than the information halflife in the soft drink industry and that the halflife in the fashion apparel industry lies somewhere in between. The prime determinant in appraising the longevity of a relative halflife is how critical the information is to a decision-making process. If the decisions must be made in seconds or minutes, the TVOI will be high. Conversely, in an industry where decisions are generally made over a period of days, TVOI will be less and the information halflife will be longer.

Technological improvements in the ability to collect and deliver information are accelerating the overall time frame within which information is perceived as having value. Thus, more rapid access to information is clearly becoming the strategic weapon of the corporate world, driven by the company's need to provide more timely services, to produce new products more rapidly, and to implement effective cost controls. The key word to gain rapid access to information is *interconnection*.

METROPOLITAN NETWORK

There are numerous types of communication networks available to users. Generally, they can be classified as private and public, and they are designed primarily for data and voice traffic, with video as a minor user. Unfortunately, there is no such entity as a unified information network, one which is efficient in carrying data, voice, and video traffic. Some of the networks we have come across most often are

1. *Common carrier network (IntraLATA):* The plain old telephone service, although after the AT&T divestiture, this network is limited to within individual geographical areas called local access and transport areas (LATA). Designed primarily for voice traffic, the network consists of circuit switches housed in local exchanges (also known as central offices), using copper loop, T carrier, microwave, and fiber optics as primary transmission mediums. Data are carried on this network predominantly through analog means. The long-term plan is to overlay this voice-oriented network with an Integrated Service Digital Network to handle data traffic more efficiently.

2. *Common carrier network (InterLATA):* Also known as the intercity network, or the long-distance telephone network. The major players are AT&T, MCI, and US Sprint, along with numerous regional carriers. The network consists mainly of digital circuit switches, using terrestrial microwave, satellite, and fiber optics as the transmission mediums. The industry trend is toward all digital networks, capable of handling both voice and data efficiently.

3. *Value added network:* A network implemented on top of the common carrier networks mentioned earlier, designed primarily to handle data; for example, Telenet's packet switching network.

4. *Local area network (LAN):* Designed primarily for data communications within a building campus environment. Data are transported over fiber optics and twisted pair or coaxial cable at high speeds, allowing personal computers, intelligent terminals, printers, and file servers to communicate with each other. Designed primarily for data transport only.

5. *Private business exchange (PBX) network:* Designed primarily for voice communications, handling phone calls within an organization entity, as well as switching circuits in and out of this entity. The primary transmission medium used is copper pairs.

6. *Private network:* Consists of a mix of facilities owned by the user as well as renting from common carriers on a full-time basis. Depending on the size of the organization, the network complexity varies. The common network components are private lines for data and voice, data modem, data and voice switches, fiber optics and microwave transmission equipment, network management computers, among others. With the introduction of T1 multiplexer functioning over high-capacity T1 circuits, the T1 private line is playing an ever-increasing role in the private network.

As mentioned earlier, to achieve a competitive edge, a company must recognize the time value of information, which can be expedited through interconnection. An organization information network, basically made up of an interconnection of the different private and public information networks, must be highly flexible and reliable because of its increase in importance. Figure 1.1 depicts a graphical interconnect relationship among these networks.

Unfortunately, not all the networks are designed as a unified information network. The intraLATA common carrier network in particular is designed for voice communications. The use of copper wire facility predominates the network. Any transport of high-speed data traffic requires the reconditioning of these local loops, and they are also the weakest link of an information network. The telephone company has recognized the local loop inadequacy and has since offered remedies such as T1 carrier service, fiber optic access, and so on. This is the beginning of a metropolitan area network. However, the availability is limited, and very often carries a special construction cost to the service. As a result, more and more organizations are heading toward a private metropolitan network.

A metropolitan network basically involves the use of terrestrial microwave or fiber optic equipment to establish communication links among facilities within a metropolitan area. The network is designed to handle both voice and data traffic—in some

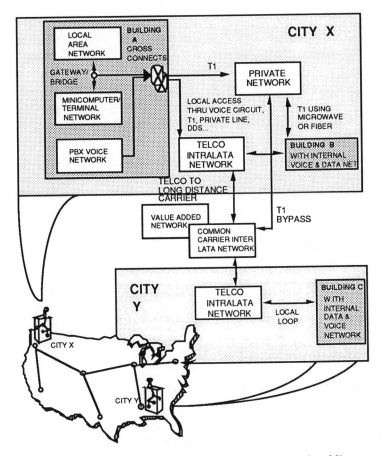

Figure 1.1.   An interconnection of various private and public networks.

cases, even full-motion video—and conforms to the telephone industry interface standard. If engineered properly, the network can provide superior connection quality and requires minimum maintenance. The network can be both private or public. In the latter case, it is a service offered by the telephone company itself (telco bypasses its own public network) or offerings by other bypass carriers.

The benefits of a metropolitan area network are

1. *Improve interconnection throughput:* The communication links are established at 1.544-Mbps (megabits per second) or higher data rates.

2. *All-digital facility:* Allows the interconnection to be done in an all-digital fashion, eliminating the need of digital-to-analog conversion.

3. *Standard interface:* Unlike the local area networks and the digital wide area networks which have multiple interfaces and protocols, some of them proprietary, the metropolitan area network adopts the long-established T1 digital hierarchy interface.

4. *Wide transport medium support:* The metropolitan network utilizes a variety of transmission mediums, ranging from T carrier over copper cable to fiber optics to terrestrial microwave. All the equipment are well established and have proven themselves under the telephone company stringent guidelines.

5. *Wide variety of interface equipment:* Whether the traffic content over a metropolitan network is voice, data, or full-motion video, there is a variety of equipment that can be used to interface the customer premises equipment to the network. This equipment includes channel bank, T1 multiplexer, and video codec, and so on.

Microwave Radio in a Metropolitan Network

Due to the advancement in radio manufacturing technology over the years, and the allocation of the 10-, 18-, and 23-GHz (gigahertz) radio spectrums for point-to-point microwave use, short-haul radios are especially reliable and cost effective for the metropolitan network environment. These radios, the majority of which are digital, transport both voice and data under the telephony digital hierarchy formats. This format is what is widely referred to as T1, T2, and T3 digital transmission speeds. The slowest transmission speed of 1.544 Mbps (T1) can handle 24 simultaneous phone calls or 24 56-Kbps data lines.

A short-haul radio system is basically composed of one or more links of terrestrial radios, transmitting and receiving at very high frequencies in the GHz range. A radio modulates an incoming T1 signal, which consists of both voice and data traffic, up to the high-frequency range and transmits them to the distant end. Figure 1.2 shows a typical arrangement, where both analog voice circuits and digital data lines are modulated up to the T1 speed by

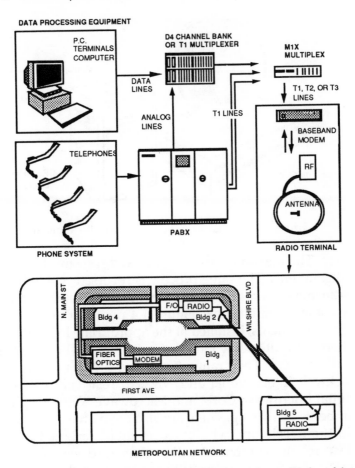

Figure 1.2 Both analog and digital signals can be multiplexed into a T1 circuit and sent over a microwave link.

means of a channel bank or T1 multiplexer. A network can then be established by setting up one or more pairs of radios. Note that each pair of radios must have direct line-of-sight from one terminal to another.

A basic digital microwave radio system has at each end a baseband unit, a radio RF unit, and an antenna (Fig. 1.3). The baseband unit, designed typically for the indoor environment, is used to modulate the incoming T1 signal up to an intermediate frequency (IF), typically 70 MHz, then transmit them via coaxial cable to the radio RF unit. The radio RF unit, designed for the outdoor environment, then modulates the IF signal further the radio

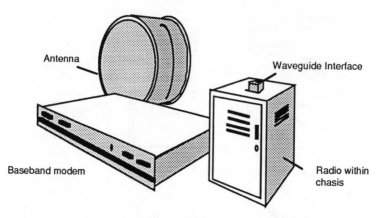

Figure 1.3. The major components of a radio terminal.

frequency. The antenna is used to aim and transmit the RF signal to the distant end. The same process is reversed when the distant radio transmits the response message, thereby establishing a full-duplex communication link. Figure 1.4 depicts the system in its separate components, installed at the appropriate locations.

There are quite a few radio manufacturers who sell the short-haul radio system. They range from long-time established giants like Rockwell International and NEC, to relatively new entrants like Digital Microwave Corp. and Microwave Networks, Inc., and the number is increasing. There are also companies, such as AT&T and Northern Telecom, who opted to OEM (original equipment manufacture) the equipment rather than to design from scratch and make their own. It is difficult to judge the radios based on the company's reputation, and the company that sells the most radios may not have the best product.

The selection criteria, of course, will be different for different customers. A 1-T1 capacity radio terminal, excluding any redundancy option or antenna, can range from $7000 to $14,000. Terminals come in different flavors, and there is certainly a difference in basic design methodology, equipment reliability, and manufacturer support.

In general, all the equipment manufacturers have the same design goal, that is, to deliver a reliable product at minimal cost. For example, the use of microelectronics components results in extremely compact and light modems, with power consumption in the 10- to 50-watt range. The outdoor weatherproof RF unit typi-

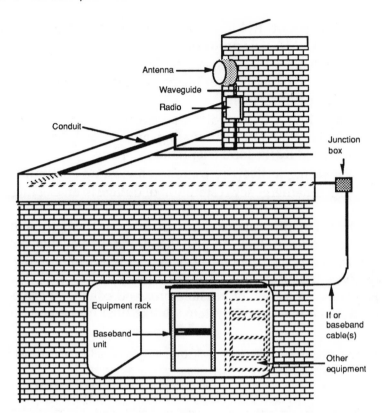

Figure 1.4. A radio terminal with its components installed at the appropriate locations.

cally measures less than 20" H x 20" W x 6" D, accompanied by a parabolic antenna ranging from 2 ft to 4 ft in diameter. The result is a system that is extremely compact, aesthetically pleasing, and last but not least, easy to install.

The applications for these radios are widespread, for example,

1. Establishing a data and voice network in a campus environment or linking several administration buildings together. In many cases, the campus environment can be as big as a major metropolitan area, where, for example, department store branches are linked back to the headquarters where the data processing center and the main private automatic branch exchange (PABX) are located.
2. Linking multiple local area networks together through T1

gateways or 10-Mbps LAN bridge adhering to the IEEE 802.3 CSMA/CD protocol.

3. Establishing a high-speed facsimile link (6 Mbps) to transmit newspaper typeset images from the layout department to the printing plant.

4. Providing a bypass direct-access link to the long-distance carrier, therefore, eliminating any special access charges incurred by the local telephone company.

5. Providing a diversity route to an existing communications link for emergency restoration of outage to an existing link.

The list of applications can go on and on. The bottom line is, a short-haul radio system can provide numerous benefits. Not only is the system very economical, it is highly efficient and reliable and gives the user direct control over the network. Table 1.1 presents a comparison of possible pros and cons for a cable-based system and a microwave system. Unlike fiber optics and T carrier

Table 1.1   Comparison Between a Microwave System and a Cable-Based System

	Microwave System	Cable-Based System
Advantages		
Highly reliable	X	X
Terminal equipment easy to install and maintain	X	X
Capable of handling data speed: In the Mbps range	X	X
In the Gbps range		X (fiber optics only)
Disadvantages		
Signal weakens due to severe rainstorm	X	
Requires procurement and lease of right of way		X
Expensive cable burial process		X
Suffers lengthy outage due to cable cut		X
Lengthy cable installation process		X

systems, terrestrial microwave does not require securing right of way; will not suffer long outages due to cable dig up, manhole fire, and so on; and is easier and less expensive to install.

However, microwave is affected by heavy rain, and this must be taken into consideration during the engineering phase. Fiber optics stands out as a superior transmission medium among the cable-based systems because of the high data rate capability; this can drive the per circuit cost down substantially. A metropolitan network seldom requires such a high level of capacity, thereby making the microwave system a more economical alternative.

The Benefit of Digital Transmission

The majority of the microwave radios used for metropolitan networks use the digital transmission method. The input and output interfaces conform to the telephone company's T1, T2, and T3 standards. Before going into what these different standards represent, we will first look into what digital transmission means and the benefit a metropolitan network has by using this transmission method.

Despite of the advance in computer technology where computers are running faster and faster, the advancement in the domestic communication network has been relatively slow in the data communication arena. Today, the most common form of data communications takes place using modems transported over the existing voice network; an alternative will be to use conditioned private lines, such as the AT&T DDS service. Unfortunately, it is universally accepted that this form of data communications can no longer carry corporate America into the twenty-first century. There are numerous limitations to these services, including quality, speed, and price. There is a basic need to upgrade the current network into a unified data and voice network.

The people who are supplying these network services, common carriers like the local telephone company, AT&T, MCI, and US Sprint, are listening. There has been a massive movement in the last few years in upgrading the quality of the network. Projects such as putting in new digital switches, installing digital microwave, and fiber optics transmission equipment all contribute to a higher-quality, more reliable communication network, a network designed to transport both data and voice traffic.

A key factor in this network migration involves the digitalization of the transmission network. Digitalization is a technique where any incoming analog signal is sampled at frequent discrete intervals. The corresponding amplitude of the signal is then represented by a sequence of numbers which may be represented in any convenient number system (such as binary) to any accuracy desired, the accuracy being determined by the number of digits used to represent each sample. A digital network transmits and processes this numerical information rather than the original analog message waveform. As the message reaches its destination, the numerical information is then converted back to its analog counterpart and is delivered to the receiving party.

For an analog system, because the transmitted signal behaves like a continuous function and can take on any value in a given duration, it is virtually impossible to filter out any baseband noise picked up along the transmission train. Any noise in the system will be reamplified as the signal. This is why data transmission over long-distance analog circuits can be a hair-pulling experience. On the other hand, digital transmission is less vulnerable to random noise introduced into the transmission train. Since there is only a limited set of numerical codes to represent the analog signal, the possibility of "guessing" the wrong number is reduced; the receiver is capable of making the correct decision about which waveform is presented even when the signal becomes contaminated by a considerable amount of noise and interference. The received signal can even be regenerated to produce a new, nondegraded waveform. Thus, the impairments introduced by the transmission facility do not accumulate unless the noise and interference are large enough to cause the line repeater to make decision errors. For the more sophisticated digital transmission equipment, countermeasures such as forward error correction schemes are implemented to reduce the errors even further. A properly designed digital transmission system will operate virtually error free.

The major advantage of digital transmission is the ruggedness of the digital signal. The limiting factor in a properly designed system is not impairments introduced by the transmission facility, but rather, the accuracy of the conversion of the original analog message waveform into digital form. This impairment, known as *quantizing noise*, may be made as small as desired by using sufficient digits in the representation of each sample value. The price paid for this ruggedness is increased bandwidth over that required for the original analog signal. However, digital signals impose different requirements on the transmission medium than do analog signals. Digital transmission terminals are also frequently cheaper to implement than are analog terminals. Therefore, there are many situations in which a digital transmission system will result in substantial economies over an analog system using the same medium.

Digital Signal Processing

The basic processing steps in transmitting an analog signal over a digital transmission system are depicted in Figure 1.5. An analog signal which has been band-limited by a low-pass filter, Figure 1.5a, is sampled at regular time intervals to produce a sequence of pulse-amplitude-modulated (PAM) samples, as shown in Figure 1.5b. Although discrete in time, the PAM samples are still analog in nature, capable of attaining an amplitude from zero to the upper and lower limits.

The PAM samples are then passed through an analog-to-digital A/D converter. The result is a set of discrete code words, each representing a snapshot of the signal at a certain point in time. The example shows a four-bit binary code word, capable of representing up to 24 discrete points. In electrical form, these binary code words can be represented by a combination of dual polarity signals, where a positive voltage represents a 1 and a negative voltage represents a 0.

The binary code generated (Fig. 1.5c) may then be translated into ternary or other multilevel digital signals acceptable to the transmission facility (Fig. 1.5d). Some of the more common ones are frequency shift keying (FSK), phase shift keying (PSK), and quadrature amplitude modulation (QAM). In this format, any noise introduced into the electrical system can be readily identified up to a certain level and filtered out, since the signal can only take on finite electrical states.

As the ternary code reaches the receiver, the binary code word is recovered (Fig. 1.5e), and a digital-to-analog converter is used to retrieve the PAM samples (Fig. 1.5f). The PAM pulses are then passed through an interpolation (reconstruction) filter to recover an approximation of the original message signal (Fig. 1.5g).

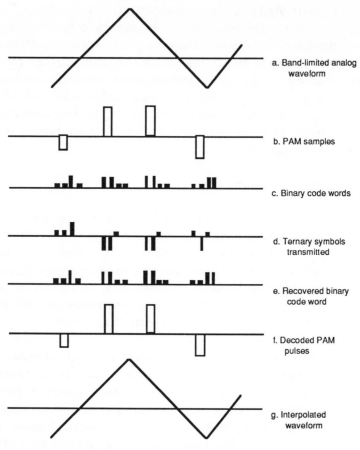

Figure 1.5.   Analog-to-digital signal processing.

The more often we sample the incoming analog waveform, and the more binary bits we use to represent the signal amplitude, the final analog waveform after an analog-to-digital and digital-to-analog conversions can reproduce the original signal more faithfully. Unfortunately, this in turn translates to more expensive converters. Furthermore, the more code words are used to represent the different analog signal amplitude, the more discrete states the transmission signal will have. This makes it harder for the receiver to determine the state of the incoming signal, especially if there is noise component in the signal. This trade-off directly affects the accurate reconstruction of the original signal. The discrepancy between the original input signal and the reconstructed output signal is termed *quantizing error*, and is the controlling signal impairment in digital transmission.

Channel Bank

In the telephony world, channel bank is the most commonly used equipment to convert an analog voice signal into a digital bit stream. It is used to modulate multiple voice circuits (e.g., from a PBX's analog ports) up to a digital bit stream of 1.544 Mbps for transmission. This high-speed digital stream, also called T1 or DS1, is the basic building block of the North America telephony digital hierarchy. In some instances, a sophisticated channel bank can even accommodate data modems at asynchronous speeds or interface with direct data input at a synchronous 56 Kbps.

Digital channel banks, or "primary pulse code modulation (PCM) multiplexers" as they are called in CCITT (International Telegraph & Telephone Consulative Committee) terminology, have two basic functions: they convert analog voice signals into digital form, and they combine the resulting digital signals into a single digital bit stream by time-division multiplexing (TDM). The actual order in which the coding and multiplexing operations are performed is a design decision which depends on the cost of the coder and the relative advantages of digital versus analog multiplexing. A typical channel bank architecture is shown in Figure 1.6. The analog voice band signal is passed through a low-pass filter which limits the signal bandwidth to less than one-half the sample frequency. The band-limited signal is then sampled at an 8-kilohertz (KHz) rate. Samples from 24 message channels are applied to a common PAM bus. During each sampling interval (125 microseconds), one sample from each channel is sequentially gated onto the bus, so that the signal on the PAM bus is a time-division-multiplexed version of all the samples, Figure 1.7. The PAM bus signal is then processed by the coder, which is shared by all the channels, to produce time-multiplexed PCM code words.

Before this PCM signal is applied to the transmission facility, some additional digital processing is necessary. First, the signaling information for each message is multiplexed with the coder output. Next, to permit correct decoding and demultiplexing of the signal at the receiver, a framing bit is inserted at the beginning of each 24-channel sequence, identifying the start of a new 125-microsecond interval, or frame. Finally, the binary signals are converted to a format accepted to the transmission facility.

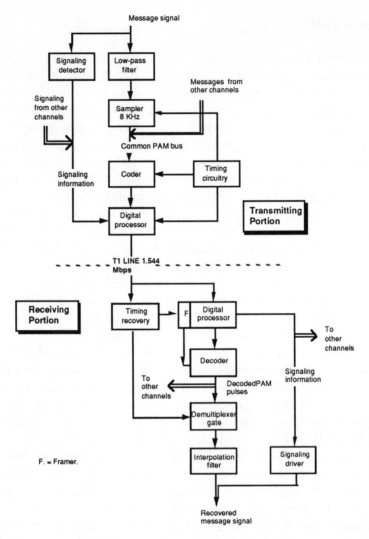

Figure 1.6.   Schematic of a channel bank.

The D4-channel bank codes each of the 24 voice band signals into an eight-bit PCM work. For five out of the six frames, all eight bits are transmitted. During every sixth frame, only seven bits are transmitted; a signaling bit for each channel is multiplexed into the eighth (the least significant) bit position of each PCM work. Finally, after the PCM word for channel 24, a framing bit is multiplexed. The framing bit varies in time according to a pattern which enables the receiving terminal to identify the beginning of the

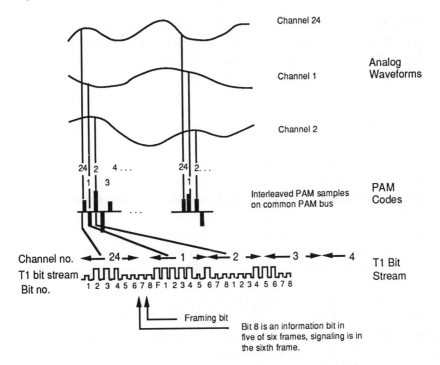

Figure 1.7. Deriving the pulse-code-modulation signal from an analog waveform.

frame and to determine whether the current frame contains signaling information. The total number of bits per frame is 193 (24 × 8 + 1); and since there are 8000 frames per second, the output bit rate is 1.544 megabits per second. In a later section, this coding scheme will be discussed in further detail.

The receiving portion of the channel bank performs the inverse functions. The incoming signal from the digital facility is first converted into binary form. Timing information is recovered so that individual bits may be identified. A framing circuit then searches the received bit stream for the framing bit pattern and synchronizes the demultiplexing circuits to the proper phase of this pattern to ensure correct interpretation of the received information and signaling bits. The signaling bits are sorted out and directed to the individual channels, and the PCM words are delivered to the decoder. In the case of D4, since only seven bits are actually used for information during a signaling frame, it is necessary to change the decoder characteristics every sixth frame for optimum performance. The output of the decoder is a series of quantized PAM pulses which are demultiplexed and applied to the individual interpolation filters. The outputs of these filters are analog signals which may be applied to the receiving voice-frequency line.

T1 Multiplexer

As the telephony industry becomes more and more digital, different services are introduced to utilize this new capability. For example, the T1 technology which was traditionally used internally by the telephone company for intercentral office transmission, is now available to the end users directly. A software-controlled T1 multiplexer was introduced to the data and voice communication arena in the mid-1980s as a result of this new service.

In essence, the T1 multiplexer provides the majority of the channel bank functions and more. It provides the user with more flexibility in managing a T1 network. Some of the major functions a T1 multiplexer provides are (Fig. 1.8)

1. Multiplexing of PCM (64-Kbps) and compressed (32-Kbps) voice circuits.
2. Multiplexing of both synchronous and asynchronous data at multiple speeds.
3. Network management functions, including automatic redundant route switchover (during time of transmission circuit failure), dynamic bandwidth allocation (according to time of day to handle variable circuit demand), and usage report generation.
4. Equipment redundancy.

The major difference between a channel bank and a T1 multiplexer basically is that the channel bank is designed to handle common voice circuits and does an excellent and cost-effective job of it. T1 multiplexer, on the other hand, offers a customer more flexibility in managing a T1 circuit and is more efficient than channel bank in handling data circuits.

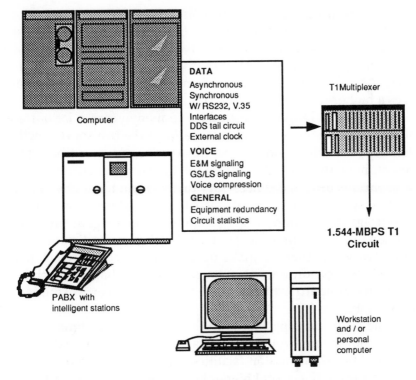

Figure 1.8.   Voice and high-speed data traffic are integrated onto a T1 link by means of a T1 multiplexer.

Digital Transport and Hierarchy

The channel bank and T1 multiplexer are two popular customer premises equipment (CPE) that can provide a T1 interface to the digital telephony network. As T1 service becomes increasingly popular, new products, especially those from the computer industry, will provide straight T1 interface. As a case in point, local area network providers are coming out with T1 gateway, allowing InterLAN communications. In the later part of 1987, computer manufacturers were announcing direct T1 ports on their machines to facilitate intermachine communications. There were also strategic alliance made between the computer makers and the T1 multiplexer vendors.

The magic behind this new T1 popularity is not due to the "invention" of T1, since T1 has been around for more than 20 years. It is the availability of T1 as a service to the customer, as well as the growing users need in both voice and data communications. T1 essentially is the fundamental building block of the North America digital communication network. Providing it directly to end users gives them exceptional flexibility in managing the network. As mentioned, digital transmission is more versatile and gives a higher-quality transport to the end user.

Because T1 has been around for so many years, a lot of transmission equipment is available to perform the transport function. Be it copper cable, coaxial cable, microwave radio, or fiber optics, they all have the proper interface for T1. To make efficient use of these various media, it is necessary to develop a hierarchy of transmission systems that operate at different information rates. A well-designed hierarchy segregates the various signal processing functions into building blocks. Each block may be flexibly interconnected with others to make transmission systems which make efficient and economical use of the available media. One such hierarchy, the standard which is used throughout North America, is shown in Figure 1.9.

The basic building blocks in the hierarchy are signal conversion terminals, multiplexers, digital processing terminals, crossconnects, and transmission facilities. All signals to be transmitted enter and leave the digital hierarchy by means of some sort of signal conversion terminal. If the signal to be transmitted is analog, A/D and D/A conversions are required. If the signal is already digital, some sort of data terminal may be required to perform further processing, such as asynchronous-to-synchronous conversion or format conversion, to make the signal compatible with the hierarchy. Multiplexers allow several lower-rate signals to be combined into one higher-rate signal to make efficient use of the available transmission facilities and also to form the interface between transmission facilities of different rates. For example, four DS1 circuits can be multiplexed up to a DS2 circuit (6.312 Mbps), seven DS2 circuits can be multiplexed up to a DS3 circuit (44.736 Mbps), and six DS3 circuits can be multiplexed up to a DS4 circuit. For the metropolitan network, typical circuit requirements are in the areas of DS1, DS2, and DS3. It is one of the reasons why the short-haul radio available for this type of voice-data transport provides capacities in those ranges.

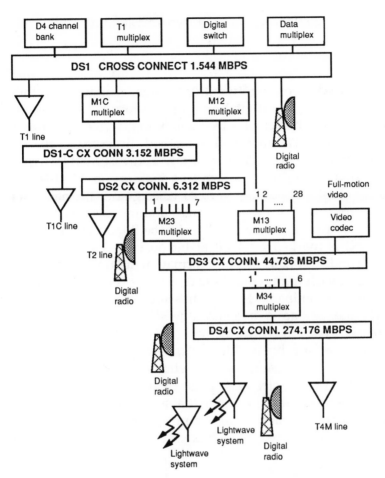

Figure 1.9.   The North America digital hierarchy.

THE DIGITAL INTERFACE STANDARDS

With the emergence of T1 as a digital communications service to end users, equipment manufacturers are introducing more and more new hardware to exploit the capabilities of the T1 pipeline fully. No longer does T1 simply mean 24 multiplexed voice channels; now there is equipment that offers compressed 32-Kbps voice, compressed video, automatic T1 switching, and network management. A metropolitan area network can certainly make use of all these features because of the popular interface. However, the key to integration is to understand truly what T1 is, what it can do, and what is it evolving to. This section will further explore in detail the different facets of the T1 transmission protocol.

The Fundamentals of T1

T1, introduced more than two decades ago, was designed to handle voice circuits in a digital multiplexed format. As a result, T1 has all the resemblance of a technology that was evolved around the voice traffic. The modulation method discussed in the previous D4 channel bank section basically addresses a T1 bit stream composition. To refresh the reader's memory, a 1.544-Mbps T1 channel is basically composed of 24 voice grade channels, each of them sampled 8000 times per second and represented by an 8-bit code. Furthermore, every 192 bits in the T1 bit stream forms a frame and is marked by 1 bit at bit position 193. Similarly, for every second, 1,544,000 bits of information are transmitted in 8000 frames, and each frame consists of 24 time slots of information.

The predominant T1 transmission protocol used in the United States utilizes a data encoding scheme known as the alternate mark inversion (AMI) method. This is basically an electrical representation of a binary bit stream, where a logical zero bit is represented by the absence of an electrical pulse and a logical 1 bit is represented by an electrical pulse of either positive or negative polarity, the only condition being that no successive 1's can have a pulse of identical polarity. For example, Figure 1.10 shows a logical bit stream translated into a bipolar electrical waveform. A vio-

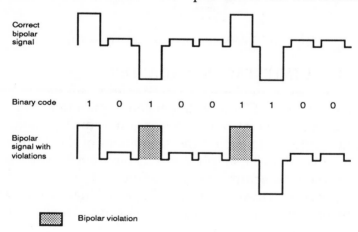

Figure 1.10. A binary code represented by a bipolar signal.

lation of the polarity rule is called bipolar violation and is defined as two successive pulses that have the same polarity and are separated by a zero level. This type of violation is often caused by noise introduced into the transmission medium. All equipment which handles AMI coding will correct any detected bipolar violation by inverting the polarity of the faulty pulse.

As mentioned earlier, AMI coding encodes spaces (zeroes) as the absence of a pulse. Network equipment will operate properly as long as there are sufficient pulses to keep them in synchronization, since a long string of zeroes will cause timing loss. Using the AMI coding as the starting point, the T1 electrical interface specification is defined in the following:

1. *Line rate of 1.544 Mbps ± 130 ppm.*
2. *Line code:* Bipolar with at least 12.5 percent of average 1's density and no more than 15 consecutive 0's.
3. *Impedence:* 100 ohms ±5 percent resistive.
4. *Pulse shape:* An isolated pulse shall fit the template shown in Figure 1.11. The pulse amplitude is between 2.4 and 3.6 volts. The pulse amplitude may be scaled by a constant factor to fit the template.
5. *Power level:* For an all-1's transmitted pattern, the power level measured at 772 KHz, within a 2-KHz bandwidth is 12.6 to 17.9 dBm. Furthermore, the power level measured in

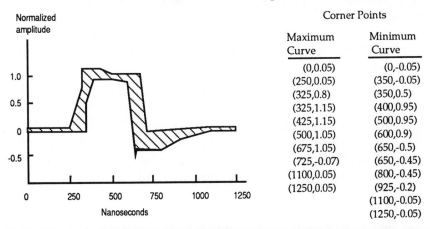

Corner Points	
Maximum Curve	Minimum Curve
(0,0.05)	(0,-0.05)
(250,0.05)	(350,-0.05)
(325,0.8)	(350,0.5)
(325,1.15)	(400,0.95)
(425,1.15)	(500,0.95)
(500,1.05)	(600,0.9)
(675,1.05)	(650,-0.5)
(725,-0.07)	(650,-0.45)
(1100,0.05)	(800,-0.45)
(1250,0.05)	(925,-0.2)
	(1100,-0.05)
	(1250,-0.05)

Figure 1.11.   T1 signal pulse shape template.

a 2-KHz band about 1544 KHz is at least 29 dB below that measured at 772 KHz.

6. *Pulse imbalance:* Ratio of power in positive and negative pulses is 0 ± 0.5 dB.

Framing Format —D4 Frame Format and Extended Superframe Format

Over the years, many changes have been introduced to the T1 protocol to ensure the proper circuit operation and improve the general network integrity. Two of the more recent changes are the D4 framing format (also known as the superframe format), and the extended superframe format (ESF).

As discussed earlier, a basic T1 channel is framed by 1 bit (called the F bit) for every 192 bits, and 1 second of transmission will transmit 8000 frames of information. These framing bits serve the purpose of identifying the frame position as well as supplying the timing information in a T1 signal. If a series of framing bits can be collected as a group and encoded with information, they can also serve the purpose of storing certain network signaling and control information. The D4 framing format (also referred to as the superframe format) basically assigns a designated "1" or "0" to each of the 12 consecutive framing bits forming a superframe. As before, the framing bits are separated by 192 bits of traffic information. The 12 framing bits are time shared to serve the following purposes:

1. *Synchronize the network terminal equipment.* This is done by the Ft bits.

2. *Identify the frames where signaling information are stored.* This is done by the Fs bits. They identify the sixth and twelfth frames where all 24 time slots within each frame are robbed of one bit for signaling purpose. In other words, a regular eight-bit time slot is robbed of one bit, the least significant bit, for network signaling purpose. This is called robbed bit signaling.

Figure 1.12 shows the D4 frame format as described earlier. A T1 stream is represented by blocks of data, each block representing one frame. The bit pulse diagram shows what is within this frame,

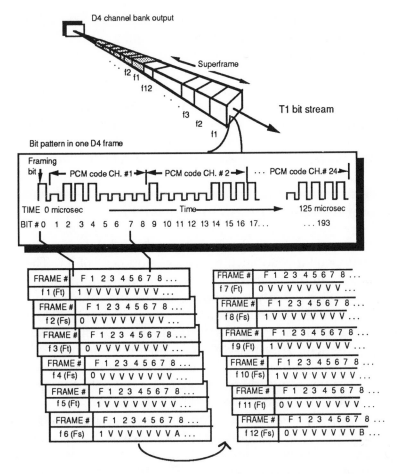

Figure 1.12.   D4 superframe format.

displaying the frame bit, followed by a series of 192 pulses. The 12 partial tables at the bottom of the diagram are the frames making up the superframe; note that each frame has a preassigned framing bit value and that the sixth and twelfth frame are the signaling frames with the last bit of each channel robbed for signaling purpose. Characters "A" and "B" represent the primary and secondary signaling bits, while character "V" represents bit used for voice coding.

To pack even more service capability into a T1 channel, providing features such as in-service circuit monitoring and trouble sectionalization and allowing rapid service restoral, a new framing format was introduced. The ESF extends the D4 format from a

12-frame structure to a 24-frame structure (193 bits/frame × 24 frames = 4632 bits) and redefines the framing bit previously used for terminal and signaling synchronization. ESF retains the terminal and signaling synchronization functions previously defined under the D4 format and introduces the cyclical redundancy check (CRC-6) code to identify one or more bits of errors within the superframe, thus allowing in service circuit performance monitoring. It also provides a maintenance data link for monitoring the integrity of the cicuit as well as providing remote alarm information.

Bipolar with Eight Zeroes Substitution (B8ZS)

The increased usage of the T1 bandwidth for data transmission has introduced a roadblock to the development of T1. The traditional alternate mark inversion code dictates a minimum "0" and "1" distribution density to preserve the timing information in the signal. However, for data transmission, it is highly likely that a long series of "0" or "1" can occur, therefore violating the AMI coding requirement. Any network termination equipment operating under the AMI coding will alter the bit stream by "correcting" the nonconforming codes, therefore changing the data content. Worst of all, there is no means of recovering this change.

To provide clear channel capability, which means an absence of restrictions on the number of successive zeroes but at an absence of restrictions on framing, a technique known as bipolar with eight zeroes substitution (B8ZS) is gradually being deployed. With B8ZS coding, each block of eight consecutive zeroes is removed, and a B8ZS code inserted in their place. If the pulse preceding the inserted code is transmitted as a positive pulse (+), the inserted code is 000+−0−+. Notice bipolar violations occur in the fourth and seventh bit positions of the inserted code. If the pulse preceding the inserted code is a negative pulse (−), the inserted code is 000−+0+−. Bipolar violations again occur in the fourth and seventh bit positions of the inserted code. At the receiving end, the terminal equipment operating under B8ZS mode will identify the bipolar violation patterns and reinsert the proper zeroes again, yet maintaining the timing information at the same time. Figure 1.13 shows the B8ZS coding scheme.

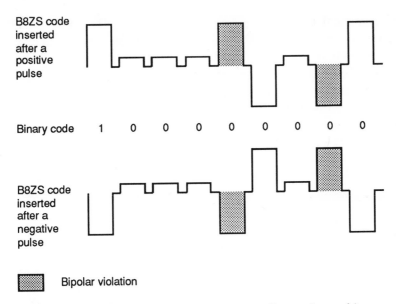

Binary code   1   0   0   0   0   0   0   0   0

▓▓ Bipolar violation

Figure 1.13. B8ZS coding is used temporarily to alter a binary stream with excessive zeroes.

Impact on a Metropolitan Microwave Network

The interface standards discussed thus far have a tremendous impact on the designing of a T1 communication network. While all the microwave transmission equipment is designed to handle the AMI coding scheme, only a portion of them are capable of handling the B8ZS coding as well. If the metropolitan microwave network is used for internal voice traffic only, AMI coding is more than sufficient to handle the requirement. However, if the microwave network is carrying traffic that requires clear channel signaling, or the network interfaces directly with other long-distance T1 network services (AT&T Accunet T1.5), furnishing the equipment with the B8ZS option is recommended.

On the other hand, the D4 framing format and the extended superframe format are transparent to the transmission equipment. Therefore, they will not affect the radio equipment selection process. D4 framing is a de facto standard that comes with all the current generation of channel banks, as well as the majority of the T1 multiplexers. ESF, because of its relatively short tenure, is not readily equipped in all T1 terminal equipment. However, it is a new generation of framing format strongly endorsed by AT&T

and the local telephone companies. Again, if the network has a chance of interfacing with any of the common carrier T1 services, the equipment selected should at least be capable of providing D4 framing, possibly even ESF, if the user sees a benefit in the additional maintenance service ESF can provide. In that case, the common carrier must be equipped to handle ESF as well.

CHAPTER 2
OBJECTIVES

Before constructing a communication system, a user must determine what objective is he or she trying to achieve before investing any dollars and effort into it. Without an objective, there is no ground to sell the proposal to upper management. Without an objective, there is no benchmark with which to compare the final completed system to see how successfully the project was carried out.

There can be multiple objectives, and they may conflict with each other. High system reliability and low cost are very often two conflicting objectives. Furthermore, the incremental gain does not necessary outweigh the incremental cost. For example, an extremely reliable system, one that is available 100 percent of the time, can cost 100 percent more than a system which works 99.99 percent of the time. Frequently, a user will establish an arbitrarily demanding performance objective, only to find out that the corresponding system cost is extremely high and unaffordable. Another user who has a low-cost objective in mind may find the completed system highly unreliable, leading to poor service quality.

Some of the possible objectives an information manager will come across include:
See Figure 2.1.

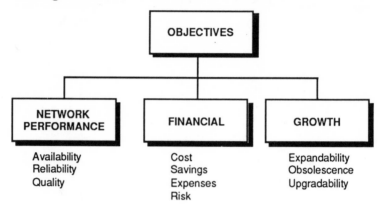

Figure 2.1. The different objectives of setting up a metropolitan network.

1. *Network design objectives,* expressed in terms of system up-time, error seconds, jitter tolerance, and so on. This will answer questions such as: How reliable should a network be? How do we specify the quality of the equipment?

2. *Financial performance objectives,* expressed in terms of a budget, net present value model, financial savings, payback, and so on. This helps to answer questions such as: Is it worthwhile to build a private network? What is the financial risk? What kind of savings will the project bring? And did the project manager do a good job in implementing the network by meeting the time and budget commitment?

3. *Growth objectives.* Financial and forecast models can be used to establish proper objectives to ensure a proper growth path for the metropolitan network. They can address issues such as equipment expandability and upgradability, as well as obsolescence.

Depending on different organizations, the foregoing objectives carry different weights. For example, a major corporation looking at cost cutting will emphasize cost savings, while another organization whose service is highly dependent on the integrity of the communication system will emphasize system reliability and operational performance. Similarly, a start-up company will emphasize growth potential while an established one may not. The important thing is to express these objectives in a manner that is understood and measurable by all parties.

NETWORK DESIGN OBJECTIVES

The network design objectives are nothing more than a list of specifications detailing the design goal of a metropolitan network. However, this is essentially the most important component of the overall objectives, as any decision made here will have a tremendous impact down the road. The major role of the network design objectives are twofold. In the beginning, it will help the orgainization to establish a realistic set of technical objectives, and afterward it will define to the participating equipment suppliers and installers a set of conditions and guidelines under which the completed system must operate.

System Reliability Objectives

The *system reliability objectives* basically define how well a network operates and whether it will operate at a bit error rate level ideal for voice and data communications. For a digital system, the parameters used for setting system objectives are

1. *Background bit error rate,* defined as the percentage number of bits in a digital bit stream that were transmitted in error with respect to the total number of bits that was transmitted in that same duration. For example, a T1 link transmitted at a data rate of 1.544 Mbps for one day and experiencing only three error bits will have a background bit error rate of

$$\frac{3 \text{ error bits}}{(1,544,000 \text{ bps} \times 60 \text{ sec} \times 60 \text{ min} \times 24 \text{ hr})}$$

   or

$$2.2 \times 10^{-11}$$

2. *Percentage error-free seconds*, defined as the percentage number of seconds within a certain time frame when one will detect no error bit. For example, during a 24-hr testing period, one bit was mistransmitted, causing an one-error second. Therefore, the error–free seconds will be 86,399 sec out of the 86,400 sec within that day, or 99.999 percent error-free second. Note that if the number of error bits is 1000 instead of 1, and they all occur within the same second, the 99.999 percent error-free second calculation is still correct. This measurement is often used for a short-term testing purpose, showing a completed system is performing at an acceptable level within, say, a 1-hr period, and is ready for customer acceptance.

    Percentage error − free seconds

    $$= \frac{\text{Percentage error} - \text{free seconds within test period}}{\text{Duration of test period (in seconds)}}$$

3. *Availability percentage*, defined as the percentage of time where a system is available. A system is said to be available when the real-time bit error rate is running under the error threshold. The error threshold for voice traffic is 1 error bit out of 1000 bits (10^{-3}), and the threshold for data traffic is 1 error bit out of 1,000,000 bits (10^{-6}). Above these thresholds, the traffic content is unusable. This availability percentage is often used as a long-term system availability objective, establishing the outage duration a user is willing to tolerate within a year.

As demonstrated, there are different ways of measuring the performance of a system. Meeting one objective does not mean the other ones are met as well. It is also the same reason why there should be multiple performance objectives, to measure the different facets of a communication system.

To establish a set of workable objectives, the user should look at what is widely available in the telecommunication industry as a reference standard, and from there, determine whether it is preferable to have an even higher standard. For example,

Service	Availability Threshold (BER)	Availability (%)	Error-Free Sec (%)
AT&T Accunet 1.5 service	$10.^{-3}$	99.70	95.00
AT&T DDS service	$10.^{-6}$	99.95	95.00
AT&T microwave short haul over 250 miles	$10.^{-3}$	99.98	NA
US Sprint last-mile T1 access	$10.^{-6}$	99.99	97.50
NYNEX wideband access (50 Kbps)	$10.^{-6}$	NA	98.75
NYNEX digital data access (56Kbps)	$10.^{-6}$	NA	99.88
NYNEX high-capacity T1 special access	$10.^{-6}$	NA	98.75

NA – Not available.

For private short-haul microwave networks, a recommended set of objectives will be

Service	Availability Threshold (BER)	Availability (%)	Error-Free Sec (%)
Private short haul, microwave—data	10^{-6}	99.97	99.95
Private short haul, microwave—voice	10^{-3}	99.97	99.95

Jitter Objective

Another network design objective involves circuit jitter. *Jitter* is defined as short–term variation of the significant instants of a digital signal from their ideal positions in time. In properly designed digital transmission systems, jitter arises from two principal sources: digital regenerators and multiplexers. Regenerator jitter is introduced by imperfections in the time recovery process, whereas multiplexer jitter is mainly related to the bit stuffing mechanism used to synchronize the low-speed incoming pulse streams. Jitter accumulation or propagation through the network is a complex nonlinear process that can be characterized conveniently only in a simplified way.

Jitter may introduce a number of impairments such as errors, outages, slips, cross-talk, and other degradations to the original signal. Thus, jitter is a particularly important parameter for the types of digital equipment connected to the network interface. The digital equipment which may be interconnected to a T1 circuit must work in the presence of jitter. Jitter characteristics are strongly dependent on the pattern of the digital signals transmitted (combination of ones and zeroes). Objectives on jitter cannot be expressed in a couple of sentences, as it involves several charts as well as the type of input signal. Common carriers such as AT&T and the Bell Telephone companies have published specifications on equipment jitter performance and should be adhered to as private network guidelines.

FINANCIAL OBJECTIVES

This objective should never be overlooked because it is the first thing upper management will look at before any capital dollars are committed to it. Furthermore, it is an excellent means of measuring how good a job the project manager is doing, as he or she is the sole person responsible for the financial aspect of the project.

Of course, the financial objectives should concentrate on minimizing cost without sacrificing quality. A thorough job must be done to identify any potential cost exposure, both immediate and recurring ones, and balance them against any potential cost saving, such as eliminating telephone private lines, increasing capacity for future growth, increasing productivity, and so on. The following outline lists some of the possible investments and savings, and depending on different scenarios, there can be others as well.

COST

1. Equipment investment
2. Installation cost
3. Outside engineering fees
4. Internal engineering/project coordination costs
5. Roof rental cost
6. Spare parts
7. Ongoing maintenance cost using outside vendor service
8. Ongoing maintenance and operations cost (internal labor)
9. Frequency coordination and ongoing protection service
10. Related legal expenditure in the real estate, permit areas
11. Travel and incidental expenses
12. Technician training

SAVINGS (TANGIBLE)

1. Telco facilities, including rate increases plus demand increases
2. Telco facilities installation charges
3. Telco facilities termination liabilities
4. Ongoing operation cost (internal labor)
5. Reduction in potential loss business due to service outage (assuming the metropolitan network offers a far superior connection than the local telephone company)
6. Tax savings due to ownership
7. Reduction in analog modem investment due to an all-digital network

SAVINGS (INTANGIBLE)

1. Improved transmission quality
2. Total digital network capability
3. Private network with total control

There are numerous ways of measuring the financial objectives, such as calculating the payback period, the internal rate of return, and the net present value. Since the limited space here prevents us from going into all of them in detail, we will just touch on one of them, the net present value model. This is considered to be the most meaningful measurement (with the fewest shortcomings) by most financial analysts. Depending on different companies' management style, a different model may be preferred.

The Net Present Value Model

The net present value model basically examines, over a period of time, the net after–tax impact on cash flow due to an investment and calculates the present value of the cash stream. For example, by installing a $50,000 microwave link, we anticipate an after-tax savings of $25,000 a year. The cash flow will then look like the one as shown in Table 2.1.

Table 2.1   Net Present Value of a Simple Cashflow

| | Year | | | | | |
	0	1	2	3	4	5
(1) Investment	–$50K	$0	$0	$0	$0	$0
(2) Annual after-tax cash saving		$25K	$25K	$25K	$25K	$25K
(3) Net cash flow (1 + 2)	–$50K	$25K	$25K	$25K	$25K	$25K
(4) Net present value (13%)	$37,930					

If the net present value is positive, the project is worth doing, as savings exceed the outlay. Notice several terms are brought up over and over again, namely,

1. *Cash flow*, a stream of cash over a time frame. Cash outflow is represented by a negative number; inflow is represented by a positive number.

2. *After-tax cash*. No matter what expenses or savings the project incurred, the government always has a cut in it. If the expenses increase, we don't pay as much tax; if the savings increase, a portion of what we save must be given back to the government. Therefore, we should be concerned only with what is left after taxes, in cash.

3. *Present value*. Since it is very difficult to compare different projects with different characteristics of cash flow over the years, a straightforward way is to look at the cash stream's present value. Due to risk and uncertainty, any dollar earned in the future is worth less than a dollar earned today, and any future earnings should be discounted. The discount formula is

$$(1 + r)^{-n}$$

where r represents the discount rate and n is the number of years into the future. The net present value of a cash stream is the sum of these discounted cash flow.

For a network project of this nature, a proper present value discount rate should be approximately 13 percent. Note that this discount rate is not based on what business the user's company is in, but rather is based on the riskiness of the project. The risk should be similar to what a company specialized in offering a similar communications network as a service will face. Referring back to the cash flow we examined in Table 2.1 and applying the discount factors, the same result of $38,000 is found. Table 2.2 the detail calculations.

Table 2.2 Calculation of the Net Present Value Using the Discount Factors

	Year					
	0	1	2	3	4	5
(1) Net cash flow	–$50K	$25K	$25K	$25K	$25K	$25K
(2) Discount factor	$(1+.13)^0$	$(1+.13)^{-1}$	$(1+.13)^{-2}$	$(1+.13)^{-3}$	$(1+.13)^{-4}$	$(1+.13)^{-5}$
(3) Discounted cash flow (1) × (2)	–$50K	22K	20K	17K	15K	14K
(4) Net present value (sum of a 5-year figures in (3))	$38K					

Net Present Value as an Objective

The present value of a cash stream is a very commonly used financial tool and is a function readily available in all the popular personal computer spreadsheet software. Table 2.3 shows a spread-sheet example of a company evaluating the feasibility of establishing a metropolitan private network to replace some aging telephone company facilities. The upfront investment is as much as $250,000 in return for five years of savings in telephone company charges—as much as $100,000 a year. However, as the example shows, these are not the only factors. Also included are other savings, expenses, tax, depreciation, and so on. The example is fairly straightforward; some of the issues worth noting are

1. *All cash flow must be after taxes.* Since any net savings means a higher gross income to a company, the tax portion must be deducted.
2. *Depreciation adjustment.* After the income tax has been taken out of the net savings figures, the capital depreciation fig-

Table 2.3   Example of Calculating a Project's Net Present Value

Description	0	1	2	3	4	5
			Year			
Investment						
Equipment investment	(200)	0	0	0	0	0
Installation of equipment	(40)	0	0	0	0	0
Engineering	(10)	0	0	0	0	0
Terminal improvement	(20)	0	0	0	0	0
Reduction in analog mode	20	0	0	0	0	0
Total capital investment	(250)	0	0	0	0	0
Periodic savings						
Telephone co. facilities	0	100	100	100	100	100
Ongoing telephone co. facilities maintenance	0	10	10	10	10	10
Total savings	0	110	110	110	110	110
Periodic expenses						
Capital depreciation of equipment	0	(50)	(50)	(50)	(50)	(50)
Roof lease	0	(5)	(5)	(5)	(5)	(5)
Network maintenance	0	(5)	(5)	(5)	(5)	(5)
Legal expenses for real estate	(5)	0	0	0	0	0
Total expenses	(5)	(60)	(60)	(60)	(60)	(60)
Net savings due to network (total savings - expenses)	(5)	50	50	50	50	50
Corporate tax on savings (38%)	(2)	(19)	(19)	(19)	(19)	(19)
After-tax net savings	(3)	31	31	31	31	31
Cash flow						
Capital investment	(250)	0	0	0	0	0
After-tax net savings	(3)	31	31	31	31	31
Depreciation adjustment	0	50	50	50	50	50
Net cash flow	(253)	81	81	81	81	81
Net present value of cash flow						
Net cash flow	(253)	81	81	81	81	81
Discount factor (13%)	1	.88	.78	.69	.61	.54
Discounted cash flow	(253)	72	63	56	50	44
Net present value	32					

ures must be added back to the after-tax net savings figures. This is because depreciation is not an actual cash flow, it is only used for accounting and tax reporting purposes.

The translation of the NPV model into budget objectives.   It is clear that the financial objective is to find a project where the net present value is positive. But once the project is identified and approved, how can the net present value objective be enforced in the day-to-day operation?   Remember, the net present value only deals with after-tax cash flow over several years, and project managers cannot be judged based on those terms. An alternative is to translate the NPV model into meaningful managerial terms such as cost savings objective, efficiency objective, and so on.

The spreadsheet example as shown in Table 2.3 identifies the different expenses and cost savings the project will have to the organization. To ensure that the net present value goal will be realized, the same figures should be reflected in the annual organization budget.

The use of NPV in satisfying the capital budgeting objective.   In reality, an organization does not have an unlimited amount of capital dollars to invest into network after network. A vice president of data processing and telecommunications is often bound by her department budget every year. If the vice president faces multiple positive NPV projects, but the department budget is just not large enough to finance all of them at once, what can she do?   This is called the capital budgeting dilema. The solution is to choose the set of projects with the *highest net present values over the same time frame, and yet the combined capital investment is within the department budget.*

For example, assume that a telecommunications department only has $5 million to invest in a private communication network that budget year, and the only two choices are

1.   A West Coast network requiring an upfront investment of $4 million with a five-year net present value of $7 million.
2.   An East Coast network having a capital requirement of $3 million with a net present value of $6 million over the same time frame.

Of course, the budget limitation prevents the department from adopting both projects, and the East Coast project seems to yield a higher return than the West Coast, with a return of $6 million to a $3 million investment. However, the rule of net present value picks the project with the highest net present value, the West Coast network, in this case. The company will wind up with a $7 million return instead of just $6 million.

GROWTH OBJECTIVES

The system growth objective is very often an overlooked objective. In a competitive marketplace where every second counts, managers spend most of their time fighting fire rather than planning for the future. If a growth objective is imposed at the beginning to cover any network need for the next three to five years, the user will reap the benefit of a steady controlled growth.

To determine the future requirement, a lot of users will take a snapshot of what is being used today and arbitrarily put a growth curve to it based on historical figures. Although this is the best information a manager will usually get, it does not consider the fact that

1. An improved network encourages more new users.
2. An improved network also encourages new usage of any excess capacity on the network since it is inherently "free" to any new user.
3. The data and voice communications arena is experiencing a dynamic growing phase, as more and more products (facsimile, T1 multiplexer, computer,) require high-speed communications.

As a result, more work is necessary to identify the potential growth. As the user explores further the type of transmission equipment available, it does not take too long to conclude that going from one level of capacity to another does not have to cost too dearly. Going from a 1-T1-capacity to a 4-T1-capacity system usually requires minor equipment upgrade. On the other hand, growing to 5 T1s or more requires the addition of a duplicate equipment or the replacement of the existing unit with an even higher-capacity unit, one with a 12-T1 capacity or 28-T1 capacity. A preferred upgrade path is one where the incremental purchases are minimal to the user.

Of course, if the growth requirement is a fairly quick one, say, growing from 1 T1 to 15 T1s in the first 12 months alone, it is better for the user to start with a 28-T1 radio right away. However, the upgrade path is not as clearcut if the growth rate is a lot slower, say, up to 15 T1s in three years. Table 2.4 compares two different ways of growing from a 1-T1 requirement up to as many as 28 T1, with minimal overall cost in mind. Only 1-T1, 4-T1, and 28-T1 radios are considered (although some vendors may have equipment with capacity in between 4 and 28 T1s). Even though both scenarios have the same estimated growth rate, the upgrade paths can be very different due to the uncertainty involved with the forecast, and how comfortable is the organization in committing upfront capital investment to long-term requirement.

When qualifying equipment vendors on future growth requirements, the best approach from a procurement standpoint is to review the estimated usage with all the equipment vendors and request a recommended equipment growth path. This should be done during the equipment comparison phase. Since any growth requirement involves investments taking place over different time periods, the net present value model discussed earlier would be extremely useful here.

Table 2.4   Growth in Demand Can Be Met by Different Equipment Growth Path

		Equipment Requirement	
Capacity Requirement	Yr	Growth Path 1	Growth Path 2
1 T1	1	Radio with 1-T1 capacity	Radio with 1-T1 capacity
2 T1–4 T1	2	Upgrade a 1-T1 radio to 4-T1 radio	Upgrade a 1-T1 radio to a 4- T1 radio
5–8 T1	3	Replace the 4-T1 radio with a 28-T1 radio	Install a second 4-T1 radio
9–28 T1	5	No action necessary	Replace existing dual set up to a 28-T1 radio

As an example, assume that a company is establishing a telemarketing department. The initial staffing level will be 40 people. Unfortunately, due to the space limitation at the headquarters, the company has to place the telemarketing department in a building a couple of miles down the road. Since the company's main PABX is located at the headquarters and the company does not want to invest in another one, it has decided to bring all the traffic back to the headquarters' main PABX. The initial voice circuit requirement between the two facilities is only 48 lines (2 T1), with a growth pattern as forecasted in Figure 2.2. There is a very high probability that the telemarketing service will be very successful, growing in double-digit rate. However, the management wants to minimize any upfront capital investment.

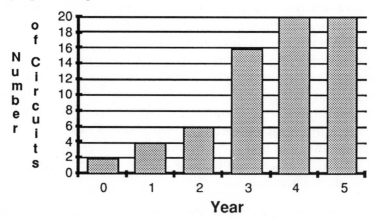

Figure 2.2.   One organization's T1 demand growth curve.

To establish the microwave link, the information manager has three options:

1. Use vendor A, purchasing a 28-T1 radio right away to handle the five-year load.

2. Use vendor A, but start with buying a 4-T1 radio. As the capacity is outgrown, replace with a 28-T1 radio. The original investment can be salvaged for $10,000.

3. Use vendor B, purchasing a 6-T1 radio but only equipped for 4 T1s. As the demand outgrows the capacity, upgrade the radio to 6 T1s. Further increase in demand beyond 6 T1s will be filled by a new 16-T1 radio, the vendor's highest capacity radio.

All three options will satisfy the demand forecast, but it is difficult to choose which one is the preferred choice. Option 3 has the lowest upfront investment, but option 1 has the lowest overall investment. Option 2 is more less an intermediate option.

Table 2.5 lays out the three investments in cash flow form. For simplicity sake, it is assumed that there is no depreciation tax benefit. According to the calculations, option 1 has a net present

Table 2.5   Comparison Between the Financial Impacts of Two Different Equipment Growth Paths

		Year				
	0	1	2	3	4	5
T1 circuit requirement	2	4	6	16	20	20
Option 1						
Vendor A						
Radio investment(28T1s)	(120k)	0k	0k	0k	0k	0k
Net investment	(120k)	0k	0k	0k	0k	0k
Net present value (13%)	(120k)					
Option 2						
Vendor A						
Radio investment (4 T1s)	(65k)	0k	10k	0k	0k	0k
Radio investment (28 T1s)	0k	0k	(120k)	0k	0k	0k
Net investment	(65k)	0	(110k)	0k	0k	0k
Net present value (13%)	(151k)					
Option 3						
Vendor B						
Radio equipment (4 T1s)	(60k)	0k	0k	0k	0k	0k
Radio upgrade to 6 T1s	0k	0k	(20k)	0k	0k	0k
Radio equipment (16 T1s)	0k	0k	0k	(110k)	0k	0k
Net investment	(60k)	0k	(20k)	(110k)	0k	0k
Net present value (13%)	(152k)					

Assumptions:
1. Vendor A's 4-T1 radio costs $65,000; its 28-T1 radio costs $120,000. Salvage value of the 4-T1 radio will be $10,000 in year 2.
2. Vendor B's 4-T1 radio costs $60,000, upgradable to 6-T1 capacity for another $20,000. Its 16-T1 radio costs $110,000 in year 3.
3. For simplicity, this example assumes there is no inflation, tax depreciation, maintenance expense, and so on.

value of negative $120,000, option 2 has negative $151,000, and option 3 has negative $152,000. The net present value rule will automatically choose option 1, the one with the least negative net present value. Unfortunately, it does not satisfy the upper management guideline of minimal upfront investment. The next choice, therefore, is option 2. Even though it is slightly more expensive than option 3 on day 1, the difference is slim. Furthermore, it provides four more T1s in capacity than does option 3.

The net present value rule helps to put the different growth scenarios in perspective. Of course, there will be circumstances where the recommended priorities can be overridden by the decision maker. It is not because the model has an error, but because some of the constraints are not considered. What the model does is to put the different options on the same reference point, time zero, and allows the decision maker to make the proper trade-off. In the example, option 2 is chosen over option 1 because the company can defer $55,000 of the investment (difference of investment in year 1) to the third year at a higher overall cost of $31,000 (the difference between the two NPV). This choice is judged to be worthwhile considering the risk of not meeting the forecast.

CHAPTER 3
EQUIPMENT SELECTION

This chapter will cover in detail the microwave equipment used for a metropolitan network. It will first introduce the overall system overview, discussing the separate functional parts of a microwave terminal. This is then followed by a discussion of the technical and subjective parameters in which a user will be interested, facilitating a thorough understanding of the different equipment selection criteria and trade-off. Finally, an equipment selection methodology called the determinant model is introduced to assist the reader in selecting the preferred set of equipment.

RADIO SYSTEM OVERVIEW

Radio Terminal

Basically, most of the 18- and 23-GHz family microwave systems available on the market today are configured similarly to each other. Each radio terminal has a baseband stage, where the input T1 signal is modulated to the intermediate-frequency (IF) stage, typically around 70 MHz. This IF signal is fed into a radio frequency stage, where the IF signal is further modulated up to the radio-frequency stage (RF), at the 10-, 18-, and 23-GHz spectrums. Along with the T1 traffic, the manufacturers may choose to transport equipment alarm signals and voice grade supervisory channels between two radio terminals by means of auxiliary channels, most often referred to as the overhead channels. Different manufacturers package their products differently. For example, for the Microwave Network's 23-GHz radio, the baseband modem is an indoor unit, separate from the outdoor RF stage. For the Rockwell Collins' 18-GHz radio, the entire system is integrated into one chassis designed for outdoor installation, and the IF stage is not readily accessible.

Figure 3.1 shows the major functional components of a radio system. At the transmitter stage, a T1 data stream is fed into the input stage of the baseband section. A cable equalizer is used to maintain the T1 data stream within its proper electrical characteristics, since this original signal may have suffered degradation reaching the input section as a result of long connecting cables. The T1 signal stream, a bipolar signal, is then converted to unipolar for processing. A modulator at the baseband stage modulates the signal up to an intermediate frequency with the help of a local oscillator. The local oscillator is responsible for supplying an extremely stable reference frequency, therefore locking the IF to its proper frequency level. At the next stage, this IF signal is modu-

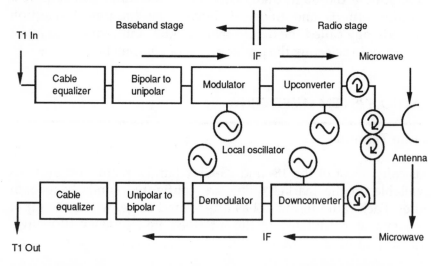

Figure 3.1.   Major functional components of a radio terminal.

lated up to the radio frequency by an upconverter referring to a second local oscillator. The output at the upconverter is the radio signal ready for transmission. It is then fed through a circulator network that splits and combines any incoming and outgoing RF signals before they reach the parabolic antenna.

The receiver stage is very similar to the transmitter stage. As the radio signal reaches the antenna, the circulator network channels the information to the downconverter. The downconverter uses a local oscillator to supply a steady reference frequency and demodulates the RF signal down to the IF stage. The IF signal is then fed through a second stage of demodulator to derive the unipolar signal, which holds the traffic information. The unipolar to bipolar conversion transforms the data stream to its proper T1 signal electrical characteristics.

Figure 3.1 is a very simplistic representation of the microwave radio. Depending on the different manufacturers, the similarity stops here. There are other important functional features to a reliable, easy-to-maintain radio system, and how they are put together depends on the individual designs. Some of these important features are the following.

Service channel. A service channel is an auxiliary channel independent of the T1 traffic, but piggyback along the T1 during the radio transmission. Frequently offered as an optional feature by the radio vendor, the service channel provides a voice and data communication link between the two radio end terminals for maintenance purpose. For example, the Rockwell Collins MDR—3x18 series radio comes with an optional plug-in service channel module to provide orderwire and fault alarm transmission capabilities. Access to the orderwire can be achieved by using a standard telephone handset with a four-wire modular phone plug. In additional, access to a 300-Hz to 16-KHz signal is provided for any fault alarm reporting systems using a regular data modem.

Alarm reporting. The more sophisticated radios usually come with complete self-diagnostic and alarm-reporting functions to facilitate the maintenance and troubleshooting process. The alarm indication can come in two modes, light-emitting diode (LED) panel display and dry contact relay closure. The first approach allows a service technician to read from the radio front panel what the equipment problem is without having to any physical measurements. Some of the LED status alarms are

1. *Power failure:* which indicates the radio power system failure.
2. *Cable fault:* which indicates open, shorted, or grounded interconnect cable.
3. *Test in progress:* which indicates that someone has put the system in test mode.
4. *BER alarm:* which indicates unacceptably high bit error rate in the radio signal.
5. *Data alarm:* which indicates the abscence of an input T-carrier signal.
6. *Transmitter output alarm:* which indicates the failure of the radio transmitter.
7. *Receiver alarm:* which indicates the absence of a radio signal from the distant-end or the failure of the near-end receiver .

The dry contact alarm approach is exceptionally useful when the radio terminal is placed at a location where a service technician is not readily available. Similar to the function of a simple burglar intrusion detection device, the dry contact relay changes its electrical state from open to ground (or vice versa), therefore triggering any remote alarm monitoring device. The minimum requirement on dry contact alarms is that it should at least provide two separate indications—major and minor alarms.

Remote loopback. The remote loopback function allows a service technician to connect the electrical baseband signal output (T1) at a remote terminal to the equipment's input port, thereby looping the circuit back to its originating end. This is an extremely useful function for circuit testing purpose.

Modem and radio connections. As referred in Figure 3.1, the baseband stage and the RF stage are separated functionally. In the case where a system is designed to have the baseband stage and the radio stage physically separated into two different equipment chassis, the information flow between the units are carried over coax cable and/or copper wires. For example, for the Microwave

Network radio, the actual IF signal carrying the traffic is passed on coax cables, while overhead control and alarm information, power, and so on are carried over regular copper pair wires. This wiring is 50all bundled together into a composite cable to facilitate installation. On the other hand, manufacturers like Digital Microwave Corporation choose to send all information, including power and local alarm information, over a pair of coaxial cables. This results in a cheaper and more effective way of connecting the baseband and radio units together, at the cost of giving up certain remote RF diagnostic functions. Figure 3.2 shows the two kinds of cables.

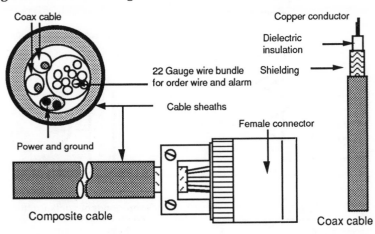

Figure 3.2. Comparison between a coaxial cable and a composite cable.

Redundancy. As an option, almost all manufacturers offer a redundancy configuration. That is, every function of the radio equipment is duplicated. The user will find two RF stages and two baseband modems, both controlled by an automatic switching controller. In the case of a component failure, the switch will redirect the traffic automatically from the primary unit to the standby unit. Of course, this feature is not free, as the cost of the terminal is more than double that of a nonredundant unit. Figure 3.3 shows a block diagram of this redundancy configuration.

It is worth noting that the redundancy feature is achieved at a trade-off of the overall radio performance. Figure 3.4 compares the protected and nonprotected radio setups. Notice that the nonprotected radio power amplifier output is fed directly into the duplexer, and out to the antenna, while the input radio signal is fed

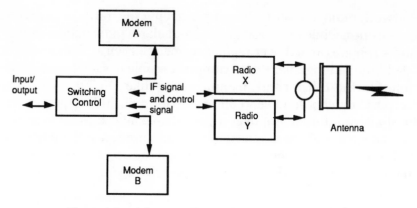

Figure 3.3. A hot-standby configured radio terminal.

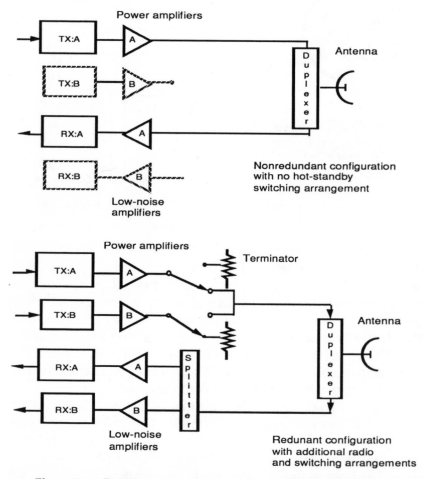

Figure 3.4. Radio terminal configured in a redundant and nonredundant mode.

directly into the low-noise amplifier (LNA) circuitry of a receiver. For the redundant setup, the output of the standby transmitter power amplifier (PA) is terminated onto a dummy load, and the output of the active transmitter power amplifier must pass through a protection switch before reaching the antenna duplexer. With this arrangement, several decibels of the available power are lost. In the receiver branch the incoming signal is divided equally between the two LNAs, resulting in another few decibels (dB) power loss. The power loss at the input and output stages typically adds up to 8 dB for the entire system.

Antenna and Waveguide System

The last stage of a radio system is the parabolic antenna. It is a passive device designed to direct the outgoing radio transmission signal to the distant end, collect the weak incoming receive signal, and send it to the radio receiver. At the same time, the antenna blocks out any interfering signal coming in at an angle with respect to the aligned antenna direction, called the *azimuth*. The Federal Communications Commision (FCC) rule specifies minimum standards on these parabolic antennas, classifying them into classes A and B. Class A antennas have a more discriminatory noise rejection envelope and are always preferred over the class B antennas. Class B antennas, however, have the advantage of lower cost and sometimes come in smaller sizes (e.g., 18 in. for 23 GHz).

There are two common ways of connecting an antenna to the radio stage. The first is through waveguide; the second is through coaxial cable. A waveguide is essentially a hollow metallic tube machined with precision to facilitate the movement of radio waves inside. The common waveguides have an elliptical cross section, and depending on the radio frequency, the cross-section area differs. As a general rule, a lower-frequency signal requires a larger cross-sectional area to facilitate wave movement. There are very stringent guidelines surrounding the use of waveguide; some of them are

1. The waveguide system should be sealed air tight to prevent any accumulation of moisture within the waveguide system. In exceptionally wet climate or long waveguide runs, it is recommended to install a dehydration system to keep the system performance up to par.

2. There should be stringent minimum bending radius

requirement to prevent kinks. Any slight dent in the waveguide induces additional signal loss and reflection.

Figure 3.5 shows a typical waveguide system with its separate components.

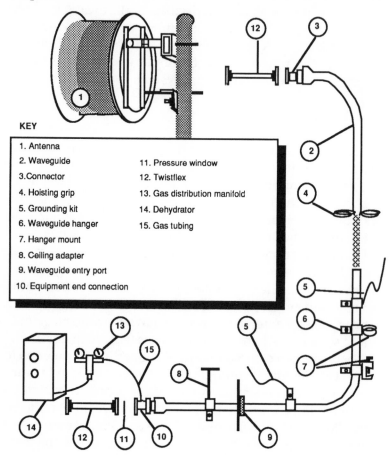

KEY

1. Antenna
2. Waveguide
3. Connector
4. Hoisting grip
5. Grounding kit
6. Waveguide hanger
7. Hanger mount
8. Ceiling adapter
9. Waveguide entry port
10. Equipment end connection
11. Pressure window
12. Twistflex
13. Gas distribution manifold
14. Dehydrator
15. Gas tubing

Figure 3.5. Different sections of an antenna and waveguide system.

The coaxial interface is more economical compared to the
waveguide interface. However, the trade-off is signal loss because
a coaxial interface has a substantially higher signal loss per foot
length penalty compared to a waveguide. As a rule of thumb, if

the distance between the antenna and the radio is more than 5 ft, a waveguide is preferred over coaxial cable. At 23 GHz, 5 ft of coaxial cable will attenuate the signal by 5 dB, compared to .5 dB with a waveguide connection. Figure 3.6 shows one of the manufacturer's coaxial interface design.

Elliptical Waveguide

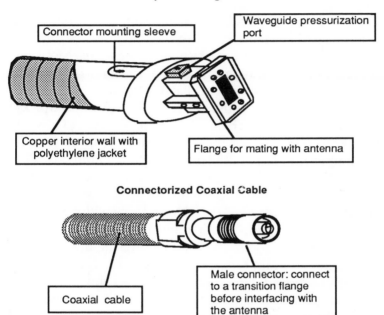

Connector mounting sleeve

Waveguide pressurization port

Copper interior wall with polyethylene jacket

Flange for mating with antenna

Connectorized Coaxial Cable

Coaxial cable

Male connector: connect to a transition flange before interfacing with the antenna

Figure 3.6. A regular waveguide flange and a coaxial antenna connector.

Once the antenna is interfaced properly to the radio, the alignment work is rather straightforward, making use of the radio transmitter and receiver. The far-end transmitter sends out a regular radio signal, while the close-end receiver shows a receive signal strength reading, typically through an automatic gain control (AGC) meter. As the antennas on both ends come into alignment, the receive signal strength increases. The process is very similar to locating metallic objects using a metal detector.

Although the antenna and waveguide system is a passive one, its importance cannot be neglected. A waveguide kink or an incorrectly aligned antenna can cause a lot of headaches, as the problem is generally the hardest to diagnose. An experienced installer is very important in this case. Any mishandling may cause gradual degradation of the system over time. For example, an improperly sealed waveguide flange allows water condensation within the waveguide, thereby causing an erratic, but very noticeable degradation of the transmit and receive signals.

Variations of a Radio System

So far, the discussion has been surrounding a 1-T1 radio. There are also radios that can handle multiple T1 circuits. The most common one is the T2 radio, which consists of 4 T1 circuits. More and more manufacturers are coming out with a T3 radio for use in the 18- and 23-GHz spectrums, which has the capacity of 28 T1 circuits. The basic radio design is the same, except that a multiplexer stage is placed before the baseband stage. The function of a multiplexer is to concentrate the T1 signals into one high-speed data channel. For example, an M12 multiplexer is used to combine 4 T1 circuits into 1 T2 circuit (6 Mbps), and an M13 multiplexer is used to combine 28 T1 circuits into 1 T3 circuit (45 Mbps). More sophisticated RF modulation schemes are used to fit these wider bandwidth signals into the FCC's allowable frequency envelope.

Another variation in the product offering involves the use of radio regenerator and repeater. Since a user very often will come across applications where there is no direct line-of-sight between the service premises, he will have to put a back-to-back terminal between the two locations to redirect the radio signal (Fig. 3.7). A repeater or regenerator replaces the two terminals and is a more economical way of redirecting the radio signal to the final destination. A repeater amplifies the incoming radio signal (as well as noise) for retransmission, while a regenerator will retransmit the incoming radio signal after reshaping the signal and eliminating any signal noise.

Evaluating this equipment is not an easy task because there are so many factors. Very often, the consideration goes beyond the theoretical system performance, cost, and sparing issues into very subjective areas such as ease of installation, customer service level, benefits of sole source. It is not the intention of this book to address all the possible factors. It will, however, address the straightforward system performance comparison area, and slightly touch on some subjective issues to prepare the reader as much as possible.

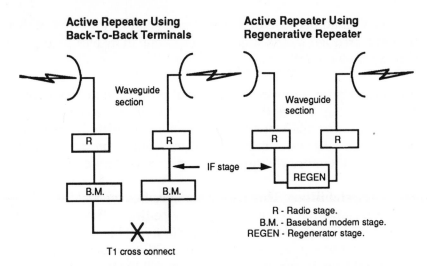

Figure 3.7.   Block diagram showing a back-to-back terminal and a regenerator configuration.

EQUIPMENT PERFORMANCE COMPARISON

Technical Parameters

As we have discussed in an earlier chapter, it is very important to come up with a set of system objectives before venturing down the engineering and implementation path. Now that we know what the final system performance requirement is (for example, an annual availability of 99.97 percent), we have to find out what is required to achieve this objective. The radio performance is a major and direct input to this process. The rest of this chapter will look at the important radio parameters that will impact the final system performance, define them, and judge them qualitatively. In subsequent chapters, we will combine them with other parameters (nonequipment specific) to derive the theoretical system performance.

The typical technical radio specifications are the following:

Operating frequency range.   This specification establishes the frequencies at which the radio is transmitting and receiving. For a short-haul radio, we are referring to the 10.5-GHz, 18-GHz, and 23-GHz spectrums. One radio designed for one spectrum will not work in another. In general, if all parameters are equal, a lower-frequency radio will perform marginally better than will a high-frequency radio because of the environmental factors. The higher the radio frequency, the more power attenuation will take place during heavy rainfall.

Transmitter output power.   The output power of a radio transmitter, immediately before the antenna and waveguide stage, is usually expressed in milliwatts or decibel milliwatts (dBm). The higher the output power, the stronger the signal will be as it reaches the distant end, therefore yielding a cleaner received message.

Frequency stability.   This parameter specifies the stability of the transmitter and the receiver frequency settings. The smaller the figure, the more stable it is. Every radio available on the market has to satisfy or surpass the minimum stability standard as set by the Federal Communications Commission. It should be noted that the FCC has established a stability standard of .003 percent for all 18- and 23-GHz equipment, considerably higher than the old .03 percent figure. As a result, it is recommended that all future equipment purchased should satisfy that standard.

Receiver sensitivity.   This specifies the minimum received signal power required at the radio antenna port to achieve a certain level of bit error rate. For example, a receive sensitivity of −83 dBm at 10^{-6} means any radio signal received at a power level of −83 dBm or higher will achieve a bit error rate of 10^{-6} or better. The lower the sensitivity figure, that is, the more negative the number is, the more sensitive is the radio and the better the receiver performs.

System gain (excluding antennas).   System gain is defined as the gain margin between the radio transmitter output and the receiver input sensitivity, or in mathematical terms

System gain =

transmitter output power − receiver sensitivity (dB)

The larger the system gain, the better overall system performance the radio system will provide. For example, if a radio has an output power of 13 dBm and a receiver sensitivity of −75 dB, the equipment system gain will then be 13 dB − (−75 dB), or 88 dB.

Mean time between failure.   Defined as the expected length of period (in years or hours) between two successive equipment failures, therefore causing a link outage, this is a theoretical value based on the individual component's predicted mean time between failure figure. For a redundant system, this figure will definitely be higher.

Dispersive fade margin.   This factor is inherent to the design of the different radios, expressed in decibels at a certain bit error rate. For example, a typical value will be 30 dB at a bit error rate of 10^{-6}. We will leave the definition of dispersive fade to Chapter 4, as it involves going into the characteristics of microwave transmission. In general, the higher the figure, the better the radio performs under signal-fading conditions.

Carrier to noise ratio.   Often expressed as carrier-to-noise ratio (decibels) at a bit error rate of 10^{-3} or 10^{-6}, the lower the figure, the better the radio performs under noisy environment. In other words, the radio has excellent rejection characteristics to outside interference.

Two of the foregoing equipment parameters, the dispersive fade margin and the carrier-to-noise ratio, although useful for comparison, are not readily available from the manufacturers. The dispersive fade margin determines how good the radio performs under signal-fading conditions, and the longer the path length, say, 10 miles or more, the more important the factor is. The carrier-to-noise ratio determines how well the radio performs under a noisy environment, especially when the user premises has other foreign systems close by using the same frequencies. In general, for a metropolitan area network (MAN), where a radio link is typically 4 to 5 miles long and has a low possibility of interference, the system performance is really determined by the first six parameters. However, in cases where a user is stretching the limits of the radios, for example, building a 15-mile link in southern California, these last two parameters should not be overlooked.

For an idea of what the typical radio parameters are, refer to Table 3.1. However, these values should not be used to judge whether a particular manufacturer's product is good or not.

Table 3.1   Typical Radio Parameters

Parameters	
Frequency	18 GHz or 23 GHz
Transmitter output power	14 dBm
Frequency stability	.003 %
Receive sensitivity at BER of 10^{-6}	−80 dBm
System gain at BER of 10^{-6}	94 dBm
Mean time between failure (Hot Standby)	80,000 hr
Dispersive fade margin	38 dB
Carrier-to-noise ratio	22

Very often, the equipment manufacturer will bring up another parameter called the modulation scheme. To transmit the T1 signal over radio frequencies, the incoming T1 signals are multiplexed into one single high-speed data stream, then modulated up to the radio frequency. There are different ways of modulating the incoming signal, hence the term "modulation scheme." Some of the typical modulation schemes for short-haul radios are frequency shift keying (FSK), phase shift keying (PSK), minimum shift keying (MSK), and digital amplitude modulation (AM). The objective of each modulation scheme is to squeeze as much information as possible within a given bandwidth allocated by the FCC while trading off performance on noise tolerance and maintaining low error transmission. Instead of going into all the detail on which modulation scheme is more superior, it is more useful to refer back to the radio parameters mentioned, as they are the results of the different modulation schemes adopted by the manufacturers.

Subjective Parameters

So far, we have discussed the important technical parameters for comparing short-haul radios. We have yet to discuss the subjective parameters. The important ones are

Ease of installation.   Even though most of the short-haul radio designs look almost identical, the installation process can be very different. For example, one vendor's equipment consists of neatly sealed units, and installing them does not require taking the units apart at all. Another vendor's equipment may be very compact, an all-in-one chassis design, but requires opening up the radio unit for fine tuning.

Power requirement. Some radios are designed to run on 110V AC power as well as –24V or –48V DC, while others may require strict –48V DC only. The latter one means an additional DC power supply is required to convert the readily available AC voltage into DC.

Special cables. Some systems may require special composite cable to link the baseband modem and the outdoor radio unit together, others may only require two coaxial cables, similar to the CATV cable. While a CATV cable costs approximately 50 cents per foot, a composite cable can cost upward of $5 per foot. The difference in cost between installing composite cable and CATV cable can be substantial if the distance between the modem and the radio is a long one.

Field tunable radio. A field tunable radio allows the user to change the radio frequency as needed. This is a very convenient feature, allowing the radio to be reused in a different setting without changing any frequency-sensitive component. It also leads to substantial savings in stocking spare parts. However, the trade-off for having this flexibility is that the frequency setting can be misaligned or mishandled by inexperienced personnel.

Spare equipment. Because of the amount of traffic riding on these metropolitan networks, the user may choose to keep a duplicate set of spares to avoid lengthy system downtime. Depending on the design of the radio, the amount of required spare parts differs substantially from one manufacturer to another. The more frequency-sensitive components a radio has, the more spare parts are required as the number of frequencies used increases.

Sole source. This concerns whether the vendor makes the entire system. For example, if a manufacturer only makes a radio with a T2 output, and OEM another vendor's multiplexer to bring the signal down to the T1 level, the entire system is really made up of multiple vendor equipment. This can be a concern if the manufacturer does not take the responsibility of testing out the entire system before delivering the different pieces to the user.

Human interface.   To service the radio equipment properly, especially in case of emergency, it is crucial to have an excellent human interface for diagnostic purpose. Alarm features like transmit/receive alarm, automatic gain control alarm, modem alarm, no input data alarm, and so on, are important. Control features like remote loopback, ordering a remote terminal to loop the signal back to facilitate overall system test, will save time and labor.

Warranty and customer support.   Issues include the length of the warranty, the type of customer support available, the size of the support organization, whether there is a local office, the repair and return policy, and the extent of the technical support a customer will get. These are very important to the ongoing operation of the network.

Design and engineering support.   Putting a radio system together takes a lot more than just powering up the radios. It involves obtaining permission from the Federal Communications Commission, designing the antenna support structure, running cables and conduits, procuring roof rights for the antenna, and so on. The more a vendor can assist, the easier it is for the customer.

Documentation.   The quality of the equipment manuals, installation procedure guide, and troubleshooting procedures are very important. Not only will a high quality set of documentation ease the design and implementation process, it will gauge how good a job the vendor is doing in the postsale support area.

Equipment upgradability.   Since the communication need will probably increase in the future, the equipment capacity today may not meet the traffic demand a couple of years down the road. The upgradability of the equipment, that is, minor alterations to increase the capacity level, becomes very important. For example, one vendor may upgrade a 1-T1 radio to a 4-T1 radio with the change of a circuit card and minimal cost, while another one does not offer this comparable upgrade but requires a change of radios.

Market position.   This is really a generic term to describe the kind of commitment a vendor is giving to this market. Just like selling consumer electronics equipment, a vendor can position itself to target at different segments of the market, hence the difference in low-cost/high-volume suppliers and high-cost/superior-quality and -service suppliers. Sometimes, the product line of a manufacturer is very limited, showing limited emphasis on high-capacity versions of the radio. Is this a sign of a weak start-up or a sign of withdrawal from the highly competitive market? Other vendors may be just OEM vendors, carrying short-haul radio to complement their other product lines. Long-term customer support is the key issue in this case. Be aware that market position does not mean market share, as a dominant supplier does not automatically translate to superior product and service with firm commitment to support their product all the way.

EQUIPMENT SELECTION METHODOLOGY

After the user has talked to the different vendors, it is time to digest the mountain of information she has in front of her. The evaluation guideline, however, should not be to choose the radio with the best technical specifications, and then hit on the vendor to get the best and final price. The proper way should be to evaluate the equipment functions and specifications to make sure they will satisfy the design objectives, meet the reliability standards, and that the system has enough capacity to handle the user's need for the next three to four years. Once a qualified set of equipment is established, go over the subjective evaluation to choose the best vendor.

One way to evaluate the subjective parameters "objectively" will be to use a weighing method called the determinant attribute model. The model's major objective is not to churn out a "right" answer based on the user's taste; instead, its purpose is to model a typical customer's evaluation process and predict what the most probable choice is. Since the buyer is selecting a product, in this case a communication system, the different product attributes are foreign and can be overwhelming. The model helps to put the different attributes in their proper perspective.

The Determinant Attribute Model

In many day-to-day buying situations, some product attributes that consumers state are important to them do not seem to function strongly in their actual brand choice. For example, many consumers will pick safety as a very important attribute when it comes to buying an automobile, yet they seldom shop around and investigate the safety level of different cars. Although safety is important, the consumers largely believe that most cars are "safe enough." Cars that appear unsafe are not considered, and the safety levels of the remaining cars are not important. The irony is that competitors normally match each other on the important attributes, and therefore the less important attributes tend to be more determinant.

To model this determinance, assume that Ms. Smith wants to decide on the vendor from whom to purchase the radio system. Table 3.2 shows five vendors (labeled A, B, C, and so on.) in her evoked set and four subjective attributes: sole source, technical support, convenience, and warranty service. To start, Ms. Smith

Table 3.2 The Determinant Attribute Model

Vendor	Sole Source	Technical Support	Convenience	Warranty Service	Dealer Preference (I)[1]	Dealer Preference (D)[2]
A	20	20	50	30	27	301.0
B	40	20	20	30	29	295.2
C	20	20	10	10	17	142.6
D	10	20	10	20	14	103.6
E	10	20	10	10	13	93.6
Importance[3]	0.40	0.30	0.20	0.10		
Variability[4]	12.25	0.00	17.32	10.00		
Determinance[5]	4.90	0.00	3.46	1.00		

[1]Dealer preference according to the importance weights is found by multiplying each vendor's attribute scores by the corresponding importance weight.
[2]Dealer preference according to the determinance scores is found by multiplying each dealer's attribute scores by the corresponding determinance scores.
[3]The attribute importance weights are assigned by the customer and must add up to 1.
[4]The variability is measured by the standard deviation of numbers in each column.
[5]The determinance is found by multiplying each importance weight by the corresponding standard deviation. A determinance score of 0 indicates a nondeterminant attribute; and the greater the determinance score, the more determinant the attribute.
Philip Kotler, *Marketing Management: Analysis, Planning, and Control,* 4/E, (c) 1980, 162 pp. Adapted by permission of Prentice Hall, Englewood Cliffs, N. J.

determines the importance of these attributes by assigning different weights (the percentages) to the subjective attributes; the percentages should total 100 percent (or 1). Ms. Smith rates sole source as most important (.40), design and engineering support next (.30), then local technical presence, or convenience (.20), and finally warranty service (.10). This selection is shown as "Importance" in Table 3.2. Table 3.2 also shows her beliefs about where each dealer performs on each of these attributes. For example, under the "sole source" attribute, vendor B has the highest score of 40, followed by vendors A and C with 20 points each, and finally vendors D and E with only 10 points each. Note that the scores in each column all add up to 100 as well.

Upon assigning the preference numbers to each attributes and vendors, the numbers should then be weighed by the "importance" factors preassigned to each attribute earlier and the weighed numbers totaled for each vendor. This is the vendor preference according to the importance weight. Which dealer will Ms. Smith prefer? If we use the importance weights, we will predict vendor B, who has a total score of 29. However, notice that the attributes differ in their variability among the vendors. For example, all five vendors are equally competent in the technical service area. To account for this variability, the model then calls for measuring the variability of each attribute using the standard deviation. The determinance of an attribute is then the product of its importance times its variability. The example shows that technical competence is a nondeterminant attribute; that is, although it is an important attribute, the customer cannot tell how one vendor excels over another in that area, and therefore has no influence over the outcome.

The final step is to multiply the determinance scores by the vendor attribute levels. The results are shown in the last column, and according to the scores, we can predict Ms. Smith will finally choose vendor A over vendor B, vendor B over C, C over D, and D over E.

Final Cost Comparison

The final step in choosing the vendor will be to look at the system cost. Having been identified, the top two to three contenders should be asked to bid on the system with their best offer. Notice this second round allows the user further to firm up his or her system objectives and configuration, weed out all the unknowns, and truly look at the bottom-line cost. It will eliminate a lot of surprises down the road and give the vendor a chance to charge extra on unanticipated work. Table 3.3 shows some possible 1987 prices on the different systems. It by no means represents an average system cost. The purpose is to give the user a feel of the investment magnitude and, most important of all, relay a sense as to what a system consists of.

Table 3.3   Cost of Four Different Radio Systems

| | System Configurations | | | |
| | Nonprotected | | Protected | |
	1-T1 Radio	4-T1 Radio	4-T1 Radio	28-T1 Radio
Equipment				
Radio terminal	$10,000	$13,000	$28,000	$40,000
2 ft. antenna	1,600	1,600	1,600	1,600
Antenna mount	500	500	500	500
IF/control cables	1,000	1,000	2,000	incl
DC power supply	incl	incl	incl	1,000
Multiplexer	0	2,500	5,000	12,000
Equipment rack	200	200	200	200
DSX patch panel	300	300	300	1,200
Installation material	200	200	200	500
Total	$13,800	$19,300	$37,800	$57,000
Sparing	$6,500	$9,000	$10,000	$12,000
Installation				
Assemble antenna mount	$1,000	$1,000	$1,000	$1,000
Assemble electronics	3,000	3,000	3,000	5,000
Test and alignment	500	1,000	1,000	1,500
Conduit system	2,000	2,000	2,000	2,000
Building permits	2,500	2,500	2,500	2,500
Total	$9,000	$9,500	$9,500	$12,000
Engineering				
Frequency coordination	$500	$500	$500	$500
Site survey	500	500	500	500
Outside engineering serv.	1,000	1,000	1,000	1,000
Internal engineering serv.	3,000	3,000	3,000	3,000
Total	$5,000	$5,000	$5,000	$5,000
Grand total	$34,300	$42,800	$62,300	$86,000

As the reader will find out, a short-haul radio system consists of a lot more than the radios. Multiplexer, power supplies, backup power, antennas mounting structure, and so on, are also part of the supporting hardware, and their cost cannot be overlooked. Also included are some figures on engineering and installation costs as a reference. The final cost, of course, will depend on the customer organization capability of handling the installation and engineering internally. Notice the installation and engineering costs as a percentage of the total system cost. For a simple 1-T1 link, it takes up as much as 40 percent of the overall cost, comparing to a mere 20 percent for a 28-T1 system. This again stresses the importance of cost control on the procurement process, and how crucial it is to have a complete list of scope of work. In Chapter 7 on implementation, we will spend more time putting the pieces together.

CHAPTER 4
SYSTEM ENGINEERING

CHARACTERISTICS OF MICROWAVE TRANSMISSION

One major criterion for designing a reliable microwave link is having a clear line-of-sight for the transmission path. Because the path of a radio beam is often referred to as line-of-sight, it is often thought of as a straight line in space from the transmitting to the receiving antenna. The fact that it is neither a sharp beam, nor is the path perfectly straight, leads to a rather involved explanation of its behavior.

By means of a parabolic antenna, the radio output signal is concentrated into a focus beam and is transmitted, through midair, to the far-end receiving station. This beam of energy tends to follow a straight line along the antenna azimuth. However, it is influenced by the atmosphere as well as the terrain and obstacles close by. Once the signal beam leaves the antenna, the signal strength starts to weaken. Rather than traveling like a straight sharp beam, the radio signal travels like the shape of a cone, with the energy getting weaker and weaker when measured at positions away from the antenna azimuth, or to the sides of the antenna. Figure 4.1 gives a graphical representation of this wave traveling characteristic.

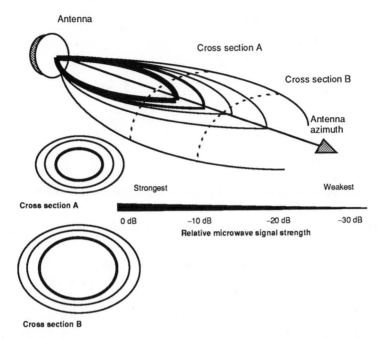

Figure 4.1. The microwave signal weakens as it radiates from the antenna.

The behavior of a microwave beam is very similar to that of light because they both belong to the electromagnetic family. The difference in their behavior is more due to the difference in the frequency. Therefore it is also useful to use light to demonstrate the propagation behavior. Some of the important characteristics are the following:

Refraction

A basic characteristic of electromagnetic energy is that it travels in a direction perpendicular to the plane of constant phase; that is, if the beam were instantaneously cut at a right angle to the direction of travel, a plane of uniform phase would be obtained. If, on the other hand, the beam enters a medium of nonuniform density and the lower portion of the beam travels through the denser portion of the medium, its velocity would be less than that of the upper portion of the beam. The plane of uniform phase would then change, and the beam would bend downward. This is called re-

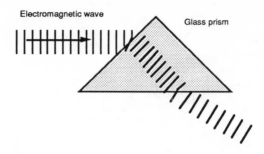

Figure 4.2. An electromagnetic wave refracts through a glass prism. Refraction occurs when the wavefront enters a new medium of different refractive index.

fraction, just as a light beam is refracted when it moves through a prism (Fig. 4.2).

The atmosphere surrounding the earth has nonuniform characteristics of temperature, pressure, and relative humidity, which are the parameters that determine the refractive index of the medium. The earth's atmosphere is therefore the refraction medium that creates beam bending and tends to make the radio horizon appear closer or farther away.

Variation in atmospheric conditions causes the transmission beam to bend, which has the same effect as the earth radius changing (Fig. 4.3). When the atmosphere is sufficiently subrefractive, the ray paths will be bent downward in such a way that the earth appears to obstruct the direct path between the transmitter and receiver, giving rise to the kind of fading called diffraction fading.

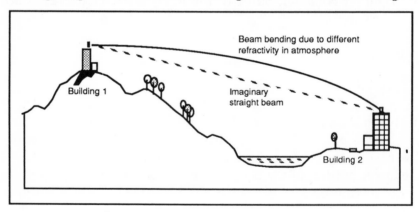

Figure 4.3. Microwave beam bending due to atmospheric refraction.

Diffraction fading of this type may be alleviated by installing antennas at a high enough level, so that the most severe ray bending will not place the receiver in the diffraction region when the effective earth radius is reduced below its normal value. This path clearance criterion will be addressed in detail later. Fortunately, for a short-haul radio designed for the metropolitan networks, the typical path distance averages 5 miles, and the extra clearance required on the antennas is minimal.

Reflection

Just as light will reflect off a mirror or shiny surface with minimal absorption, so will microwave reflect off large smooth surfaces. A large body of water is to microwave as a mirror is to light. Microwave reaching a body of water will bounce off the surface with a 180 degree change in signal phase. If this secondary signal reaches the receiver, it will cancel the primary signal. Although there are ways to overcome reflection, for example, by means of shifting the relative heights of the antennas at both ends to reposition the reflection point, or using a space diversity antenna system—the use of two separate antennas at each end, allowing the radio to switch from one to another if the signal gets too weak on one of them—these countermeasures are expensive and are generally reserved only for the more sophisticated long-haul microwave system. As a rule of thumb, any radio transmission over a lake, reservoir, or river that covers more than a mile should be planned with caution and avoided if possible.

Figure 4.4 explains this reflection phenomenon graphically. The reflection point is dependent on the relative elevation of the two transmission points with each other and with the reflection surface, as well as the distance between the two end points. To find out whether a water body will potentially reflect the microwave signal and cause interference is relatively easy: first, establish a vertical profile of the terrain between the radio transmitter and receiver; then draw a secondary beam linking the building at one end to a mirrored image of the building at the opposite end. If this secondary beam cuts across any water surface along the path profile, there is a reflection problem.

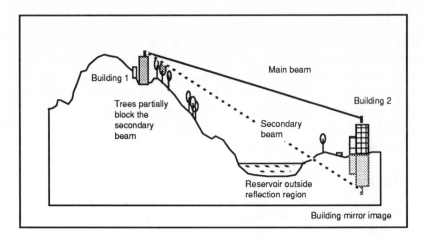

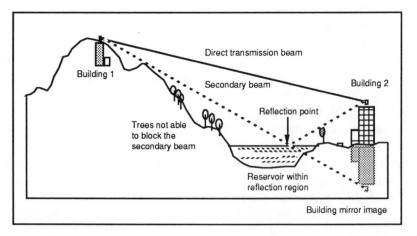

Figure 4.4. Possible signal reflection as the result of a large water body.

Multipath Fading (Dispersive Fading) Due to Stratified Atmosphere

Under normal conditions there should be only one propagation path between the two antennas on a line-of-sight radio relay link. In practice, there are occasions when more than one propagation path exists, and interference among the signals received over these paths may give rise to significant fading. These additional paths are generally due to reflections either from the ground and surface structures or from multiple tropospheric layers with rapid vertical variations causing discontinuities in the atmospheric refractive index.

During stratified atmospheric conditions on a line-of-sight radio relay system, various height configurations between the transmitting and receiving antennas can cause multiple fading, also called dispersive fading. It exhibits a frequency and duration of fading which are related to the variation of the structure of the refractive index in time. On overland paths and in wet climates, these variations normally occur during the night and early morning hours of summer days.

The countermeasures to dispersive fading are inherent in the radio system. Equalizing circuits such as slope adaptive equalizer and transversal equalizer are used in the more sophisticated longhaul systems. For the short-haul radio, because the average length of a path is relatively short, these sophisticated equalizers are not used. This is also the reason to watch out when using the short-haul radio for long hops, say, 10 to 15 miles, since the defense mechanism is not built in.

Obstruction

Microwave cannot penetrate any solid objects. That is why there is a line-of-sight requirement for the transmission path. By line-of-sight, we mean there cannot be anything along the path, not trees or a moving jumbo jet. When determining line-of-sight during the winter time, be careful to take into consideration foliage blockage during the warm seasons. Furthermore, allow 10 to 20 ft for future growth. If the transmission path is right above a major airport, blockage due to jets can cause signal fading to the extent of outage. On the other hand, because the microwave beam has a diameter of 10 to 20 ft in midpath, small obstacles like light poles, electrical cables, and so on, have a negligible effect on the system performance.

Like light, microwave can penetrate windows as well, as long as the glass has no heavy metal content. Best of all, microwave can even penetrate fiberglass although it suffers a loss of power. This means that in cases where antenna placement present an aesthetic problem to the user, it can be placed indoors, shooting through a window, or placed on the rooftop, surrounded by a fiberglass facade disguised as a small penthouse. The loss in signal power due to a glass window or fiberglass facade should be taken into consideration during the system design phase.

Rain

Precipitation, especially rain, causes absorption and scattering of a radio wave. These effects combine to produce attenuation. Although all frequencies are subject to these effects, attenuation is of practical importance only for frequencies above 10 GHz and for percentages of time for which there is heavy precipitation. For short-haul radio in the 18- and 23-GHz range, attenuation due to rain is a significant factor in the overall system performance. Snow and fog, on the other hand, have negligible impacts at these frequencies.

Numerous research has been done on the effect of precipitation on microwave. There is an established relationship between microwave signal attenuation and rain under parameters such as the microwave frequency, the signal polarity, the rain droplet size and shape, the size of the rain cell, and the rain rate.

For an 18-GHz or 23-GHz short-haul radio, this effect dominates the system performance considerations. You can minimize this by increasing the effective power of the transmission signal. Also worth noting is that during rain, the atmosphere is saturated with moisture and is no longer stratified. In other words, dispersive fading and rain fading are two independent events exclusive of each other, and each has its share of expected outage time.

MICROWAVE PATH ENGINEERING

From a system design standpoint, all the microwave characteristics just noted should be taken into consideration. Enough research and field measurements were performed to come up with the empirical models to estimate the performance of a radio system. The rest of this chapter will concentrate on the different design models showing the important steps in designing a reliable system.

The basic objective of route design is very straightforward: to link a user's terminal to its destination with the least cost and yet satisfy the design performance objective. In a metropolitan area, it is very common to find out that although the two end points are merely a short distance apart, the line-of-sight is blocked because of another structure in between; therefore, the alternative is to link the two points using an intermediate repeater. Notice as the number of repeater points increases, the associated equipment increases, and so does the cost of maintenance and real estate rents. In a large city like Los Angeles where a user's facilities can be very far apart, additional repeaters are required to regenerate the signal to maintain an acceptable system performance. This whole route design process can therefore be very cumbersome.

The necessary steps to ensure an accurate and reliable path are to

1. Establish path profile.
2. Establish clearance from any potential obstacles.
4. Check for potential reflection point.
3. Establish basic system configuration and parameters.
4. System performance calculations.
5. Site survey to assure line-of-sight.

The steps are listed in their order of preference to minimize wasted engineering time, although it is perfectly acceptable for a designer to choose one step ahead of another. For example, one may do a site survey first because one is situated in a particular building already, instead of spending unnecessary time on plotting path profile. Common sense should be the key to prioritizing the different steps. Very often, especially for large corporations, the facilities cover different major metropolitan areas, and it is difficult for the designer to cover all the ground at the same time.

Establish Path Profile

The objective to establish a path profile along the transmission path is to ensure sufficient obstacle clearance. Maps are the principal source of data, both for office study, which usually precedes the field survey, and for the field survey itself. There are several types of maps which will be helpful to the profile work. The best ones are the topographical maps published by the U.S. Geological Survey, printed on a scale of 1:24,000, covering an area of 7.5 minutes in both longitude and latitude (1 in. = 1/2 mile). These maps

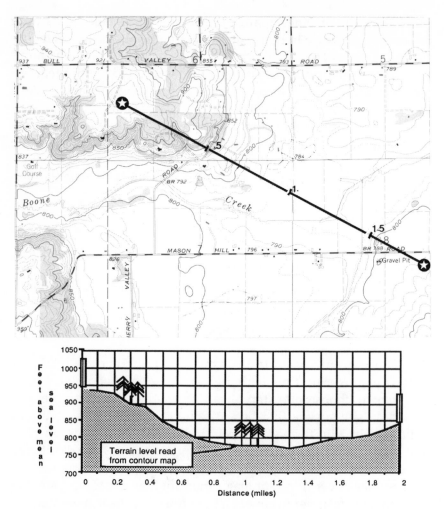

Figure 4.5. Deriving a path profile from the geological contour map.

show the terrain contour as well as the vegetation in the area. Large buildings are shown as well. However, there is no indication of any structure heights. Depending on the location of the terminals, more than one map may be required to show the entire path.

Upon identifying the end terminals on the map(s), a straight line representing the transmission path is drawn between the two points. Figure 4.5 shows the transmission path overlapping the contour lines. Follow along this transmission path, taking down terrain elevation data at regular intervals. The set of data (ground elevation measured in feet versus distance measured in miles) can then be plotted graphically, giving a terrain profile.

At the same time, the designer should look for bodies of water that overlap a mile or more with the transmission path and look for the point of reflection in relation to the two end points (as discussed earlier in this chapter). If there is a potential problem with reflection, an in-depth investigation is required.

Since the contour map does not show the height of the vegetation and structures on the ground, it is useful to assume obstacles of 60 ft on the ground level in the beginning. If the transmission beam passes this first assumption, the chance of getting a clean line-of-sight is higher.

Establish Path Clearance

Thus far, all the discussions have been based on a flat earth and straight microwave beam, while in fact neither condition is true. The relative curvature of the earth and the microwave beam is an important factor when plotting a profile chart. Although the surface of the earth is curved, a beam of microwave energy tends to travel in a straight line under normal conditions, with occasions of bending downward by a slight amount due to atmospheric refraction. The amount of bending depends on atmospheric conditions, and the degree and direction of bending can be conveniently defined by an equivalent earth radius factor, K. This factor, K, multiplied by the actual earth radius, r, is the radius of a fictitious earth curve. The curve is equivalent to the relative curvature of the microwave beam with respect to the curvature of the actual earth minus the curvature of the actual beam of microwave energy. Any change in the amount of beam bending caused by atmospheric conditions can then be expressed as a change in factor K. This relative curvature can be shown graphically, whether as a curved earth with radius Kr and a straight-line microwave beam or as a flat earth with a bent microwave beam having a curvature of Kr. Both representations are depicted graphically in Figure 4.6. From a practicality and ease-of-use standpoint, the second method—flat earth with a bent microwave beam—is preferred.

A path profile plotted on rectangular graph paper with no earth curvature and with the microwave beam drawn as a straight

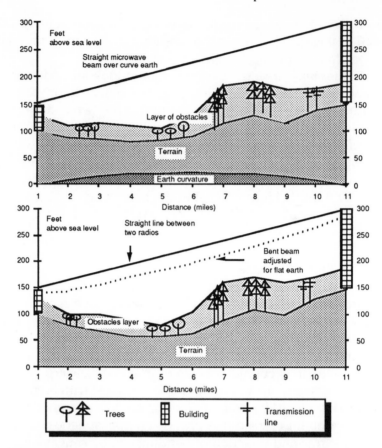

Figure 4.6. Two ways of representing earth curvature. Upper panel shows a curved earth surface with a straight beam; lower panel shows a flat earth with a bent beam.

line between the antennas represents conditions when the beam has a curvature identical to that of the earth. The K factor equals infinity at this point, which represents one of the extreme conditions. To complete a propagation study, it is necessary to show the path of the beam for other expected values of K. Under normal atmospheric conditions, the K factor is equal to 4/3. Another more stringent criterion is when K equals 2/3. Figure 4.7 shows three different scenarios:

1. K equals infinity. This is a case when the fictitious earth curvature is infinity, that is one of flat earth. The distance be-

tween a horizontal beam and the earth surface is always constant.

2. K equals 4/3. This is found to be the normal transmission condition. The effective earth radius is larger than the actual earth radius and therefore gives the signal beam a farther reach than if it is transmitted in the vacuum.

3. K equals 2/3. This effective earth radius is smaller than the real earth radius. A rare and stringent condition used for long-route-planning purposes.

4. K equals 1. This condition is not shown in Figure 4.7; it represents the case where the fictitious earth curvature is actually the earth curvature. If this is shown in Figure 4.7, common sense will tell us that the beam between the two end points will be a straight one.

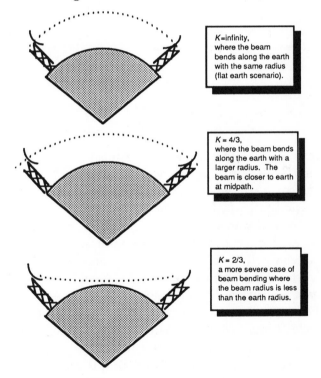

Figure 4.7.   Different effective beam bending as a result of *K* factors.

The curvature for various values of K can be calculated from the following relationship:

$$h = \frac{5 \times 5}{1.5 \times 4/3} = 12.5 \text{ ft.}$$

where

h = the change in vertical distance from a horizontal reference line, in feet

d_1 = the distance in miles from a point along the path to one end

d_2 = the distance from the same point to the opposite end of the path, in miles

K = the equivalent earth radius factor

For example, assume a normal atmospheric condition where K equals 4/3 and the path distance is 10 miles long, at the middle of the path:

$$h = \frac{d_1 \times d_2}{1.5 \times K}$$

In other words, the microwave beam is really 12.5 ft closer to the earth surface than what we would have predicted if we assume a flat earth. If the K factor is more severe, say, at 2/3, h would be equal to 25 ft. It is obvious to conclude that the longer the transmission distance, the more important this adjustment is because the earth curvature is more pronounced.

So are we ready to say that once the center of the microwave main beam clears the obstacle, with adjustment to the earth curvature, we have a clear path? Unfortunately, the answer is no. As we discussed earlier, microwave is not really a sharp beam, it is a dispersion of electromagnetic waves in a conical shape. As we go farther and farther away from the transmitting antenna, the base of the cone increases. At least the majority of this conical beam has to clear the obstacle, and the objective measurement is defined by a term called Fresnel zone.

Figure 4.8 shows a transmission beam going from point A to point B, over an obstacle point located in between. Assume that

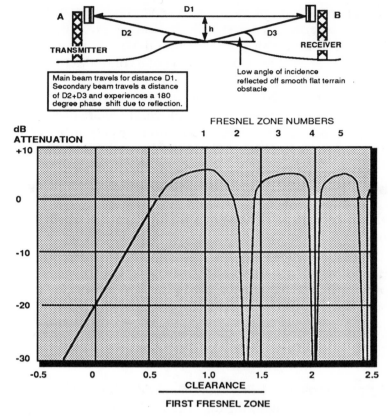

Figure 4.8.   Derivation of Fresnel zones.

this obstacle has a smooth surface, for example, a dry lake bed at high elevation. Part of the microwave energy will bounce off this surface at a low angle of incidence and reflect back to the receiver, like the effect of a mirror, with a 180 degree shift in phase. Notice that this second beam takes a longer route home, and depending on the distance of the obstacle height in relationship to the main beam (the distance h), the second beam can enhance or attenuate the original signal by the time it reaches the receiver.

To understand this further, let's adjust the clearance height h and observe the signal behavior at the receiver end. At h equals 0 ft, most of the signal is blocked by the obstacle, therefore giving only a very weak signal at the receiver end. As the clearance h increases from zero, the signal strength at the receiver will increase. The signal will peak at a point where the primary and secondary transmissions are exactly in phase at the receiver end; the parameter h at this point is termed the first Fresnel distance. An imaginary ellipsoid centered on the signal path azimuth with a distance of h is called the first Fresnel zone. If the value of h continues to increase, that is, the clearance continues to increase, the secondary transmission will begin to take longer to reach the receiver, causing an out-of-phase condition. The received signal will therefore decrease immediately beyond the first Fresnel zone. As the clearance h continues to increase, the in-phase and out-of-phase conditions will continue, resulting in peaks and valleys in the transmission signal strength at the receiver end. The individual peaks and valleys mark the subsequent Fresnel zones. The clearance criterion dictates the obstacle to be clearing at least 60 percent of the first Fresnel zone. But the million-dollar question is: How do you determine the size of the Fresnel zone?

It is quite easily derived that (providing the reader took college physics in electromagnetic wave) the first Fresnel zone has a radius of

$$F_1 = 72.1 \times \left(\frac{d_1 \times d_2}{F_{\text{GHz}} \times D_{\text{mile}}} \right)^{1/2}$$

As can be seen, the Fresnel zone is dependent on three factors: the total path length, the distance of the location in question with respect to the two end points, and of course, the frequency (which determines the wavelength). For a 23-GHz system over a 10-mile path, the Fresnel zone at midpath is going to be approximately 24 ft. Using the 60 percent criterion, Fresnel zone clearance should be at least 15 ft above the obstacle.

Now that we have considered both the earth curvature factor and the Fresnel zone clearance criterion, we are ready to set a recommended path clearance criterion:

For a heavy route with high reliability system objective, the criteria are:

At least 0.3 F_1 at $K = 2/3$ and at least 1.0 F_1 at $K = 4/3$,

whichever requires the greater height.

Table 4.1   Obstacle Clearance Criteria over Various Path Lengths (miles)

Path Distance	1.0	2.0	3.0	4.0	5.0	6.0	7.0	8.0	9.0	10.0
F_1 (10 GHz)	11.1	15.7	19.3	22.3	24.9	27.3	29.4	31.5	33.4	35.2
F_1 (18 GHz)	8.5	12.0	14.7	17.0	19.0	20.8	22.5	24.0	25.5	26.9
F_1 (23 GHz)	7.5	10.6	13.0	15.0	16.8	18.4	19.9	21.3	22.6	23.8
Earth Curvature										
$K = 1$	.2	.7	1.5	2.7	4.2	6.0	8.2	10.7	13.5	16.7
MPC 10 GHz	16.9	20.1	23.1	26.1	29.1	32.4	35.8	39.6	43.5	47.8
MPC 18 GHz	15.3	17.9	20.3	22.9	25.6	28.48	31.7	35.1	38.8	42.8
MPC 23 GHz	14.7	17.1	19.3	21.7	24.3	27.0	30.1	33.5	37.1	41.0

MPC: Midpath clearance based on the criteria of $0.6F_1 + 10\text{ft}$ at $K = 1$

Both criteria must be evaluated along the entire path. For light route, which is applicable to most short-haul radio links or medium reliability systems

$$\text{at least } 0.6F_1 + 10 \text{ ft at } K = 1$$

Table 4.1 shows the midpath clearance criterion for a light route using several path lengths. The clearance required for short path is negligible while that of a 10-mile path is as much as 47.8 ft for a 10-GHz system. In a metropolitan network environment where the path length is usually less than 10 miles, it is safe to adopt a general rule of thumb of using a total clearance distance of 45 ft above any obstacle, about the height of three stories of a commercial high-rise building.

Check for Potential Reflection Surface

The last part of establishing a reliable path involves looking for any potential signal reflection surface. If the contour map shows any body of water (1 mile or wider) situated right underneath the transmission path, the exact reflection point must be determined. As discussed before, the reflection point is dependent on the relative heights of the two transmitters with respect to the reflecting surface and the distance between the two end terminals. When K equals infinity, that is, assuming a flat earth condition, the relation between the antenna heights and the distances from the respective ends to the reflection point can be expressed by

$$\frac{h_1}{h_1 + h_2} = \frac{d_1}{d_1 + d_2} = \frac{d_1}{D}$$

where

h_1 = he elevation of the lower antenna in feet above the reflection surface

h_2 = the elevation of the higher antenna in feet above the reflection surface

d_1 = the distance in miles from the h_1 end to the reflection point

d_2 = the distance in miles from the h_2 end to the reflection point

$D = d_1 + d_2$

Rewriting the formula, we have:

$$d_1 = nD$$

where

$$n = \frac{h_1}{(h_1 + h_2)}$$

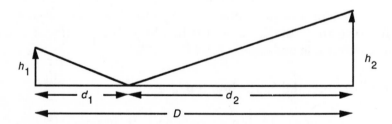

h_1 = elevation of the lower antenna above the reflection point

h_2 = elevation of the higher antenna above the reflection point

D = distance between the two antennas

d_1 = distance of reflection point to the h_1 location

$\quad = D[h_1 / (h_1 + h_2)]$

d_2 = distance of reflection point to the h_2 location

$\quad = D - d_1$

Figure 4.9. Determining the reflection point between two terminals. K = infinity.

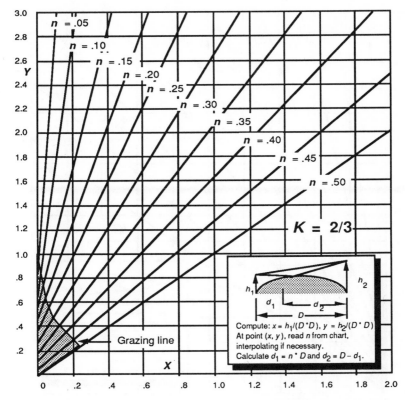

Figure 4.10.   Point of reflection of overwater microwave path. $K = 2/3$.

Figure 4.9 shows a graphical representation of this relationship.

Unfortunately for long paths, the formula is not sufficient to pinpoint the most likely reflection point. As discussed before, the microwave beam-bending behavior and the earth curvature are factors with which the designer must contend. For values of K other than infinity and for unequal relative antenna elevations, the geometric relation involves cubic equations whose solution is very cumbersome. Fortunately, graphical solutions have been worked out to locate the reflection point relatively easily.

Adopt the same $d_1 = nD$ relationship mentioned in the flat earth case, except now n is a constant read off from Figures 4.10 (for $K = 2/3$) and 4.11 (for $K = 4/3$). To find n, we must first determine X and Y, where

$$X = \frac{h_1}{D^2} \qquad Y = \frac{h_2}{D^2}$$

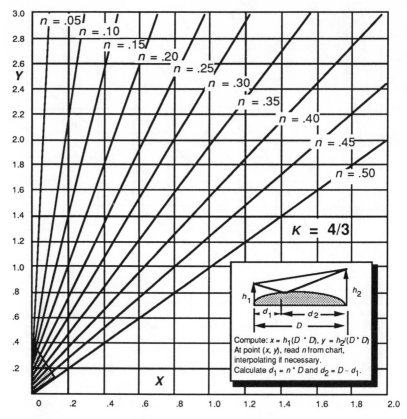

Figure 4.11. Point of reflection on overwater microwave path. $K = 4/3$.

n is then read off the chart for point (X, Y), interpolating if necessary. Once we find d_1, d_2 is simply $D - d_1$.

The graphical solution provides accuracy adequate for most
work. If a more accurate result is desired, the following two rela-
tionships are useful:

$$\frac{h_1}{d_1} - d_1 = \frac{h_2}{d_2} - d_2 \qquad (K = 2/3)$$

$$\frac{h_1}{d_1} - \frac{d_1}{2} = \frac{h_2}{d_2} - \frac{d_2}{2} \qquad (K = 4/3)$$

The graphical results should be applied to the appropriate
formula. If the reflection point is correct, the left side of the formula will be equal to the right side. If the two sides are not equal,

d_1 should be increased by a small amount while d_2 should be decreased by the same amount (the condition $d_1 + d_2 = D$ must be met at all time). If the difference increases, try decreasing d_1 instead. Once both sides of the formula match, the reflection point is found.

The small shaded area in the lower left-hand corner of each figure represents conditions for which the path would be below grazing and a true reflection point would not exist.

Establish Basic System Configuration

Now that we have determined a line-of-sight path, the next step is to determine the system configuration. By now, the user should have collected specifications from different radio manufacturers and is familiar with the parameters mentioned in Chapter 3. Based on the specific radio and path parameters, we can build the first system.

Figure 4.12 is a block diagram showing the different stages of a radio system. The radio signal generated by the radio has to go through a series of connecting equipment, which includes waveguide, branching network, and parabolic antenna before it reaches the air medium. Each of these stages affects the power of the signal. Over air, the radio signal suffers further alteration, which is directly proportional to the logarithmic path distance. As the radio

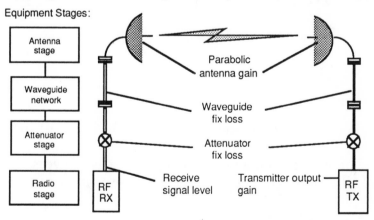

Figure 4.12.   Different stages of a radio terminal.

signal reaches the receiving terminal, it goes through the antenna and waveguide stages again before reaching the receiver section.

The goal is to assemble a system so that the receive signal level, the signal power level at the input stage of the radio receiver, is 30 dB or higher than the receiver sensitivity at the threshold of 10^{-6} bit error rate. In other words, we are allowing a fade margin of 30 dB for the entire system so that the signal can fade for 30 dB, and we still have a performance of 10^{-6} or better. Note that the 30-dB fade margin is a reference figure for the preliminary route design phase only; we will address the system performance in relation to fade margin in detail in the next section.

To determine the receive signal level, we simply add up (or subtract) all the power-related components from one end of the radio terminal to the other. For example, referring back to Figure 4.12 again, assume the radio transmitter has an output power of 20 mWatts, or 13 dBm, and the signal passes through a waveguide network. According to the waveguide manufacturer, this signal will suffer an attenuation of 1.5 dB when going through the network. Therefore, the net signal power will be at 11.5 dB at the back of the antenna (refer to the Appendix for a discussion of decibels).

Unfortunately, calculating the loss from one antenna to another is not as straightforward. According to the study on electromagnetic wave, radio wave attenuation in free space (free space loss) between two isotropic antennas (omnidirectional type) can be approximated by the function

$$-A_{\mathrm{dB}} = 96.6 + 20 \log_{10} F_{\mathrm{GHz}} + 20 \log_{10} D_{\mathrm{miles}}$$

where

F_{GHz} = signal frequency

D_{miles} = distance between the two antennas

For example, for an 18-GHz signal travelling over a distance of 10 miles, the signal loss is −141.8 dB. For a distance of 1 mile, the loss will be −121.7 dB.

In a metropolitan network, the distance between two points can be very close, sometimes as short as a few hundred feet. The formula is still a reasonable approximation. Another way of ex-

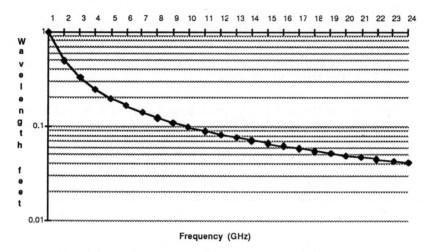

Figure 4.13.   Relationship between frequency and wavelength.

pressing the free space attenuation over a very short distance, such as a couple of hundred feet, can be stated as "for a distance equal to 1 wavelength, the loss is 22 dB, and each time the distance is doubled, another 6-dB loss is incurred." For example, at 18 GHz (where wavelength equals .055 ft), the loss will be 22 dB at .055 ft, and 28 dB at .11 ft. Expressing this relationship mathematically can bring us back to the A_{dB} formula discussed earlier. Figure 4.13 shows the relationship between frequency and wavelength, where wavelength (in feet) equals 983,600 divided by the frequency.

With the help of the parabolic antennas, the signal energy can be concentrated between the two antennas, and therefore there is a gain involved compared with using omnidirectional antennas. The gain factor for each parabolic antenna should be added to the free space loss figure to determine the net path loss. The gain factor for parabolic antennas over isotropic antennas is determined by the signal frequency and the diameter of the dish. A good approximation is

$$G_{dB} = 20 \log_{10} B_{ft} + 20 \log_{10} F_{GHz} + 7.5$$

$$B_{ft} = \text{Antenna diameter in feet}$$

The gain figure is often published by the antenna manufacturer and should always be the preferred figure to use when working on path calculations. Table 4.2 shows some typical gain figures. Referring back to the example, the net path loss of the signal from one 4-ft antenna to another is

$$G_{dB} \text{ of antenna } 1 + A_{dB} \text{ path loss} + G_{dB} \text{ of antenna } 2$$

or

$$44.7 - 141.8 \text{ dB} + 44.7 \text{ dB} = -52.4 \text{ dB}$$

Since the signal level at the input stage of the first antenna

Table 4.2 Antenna Gains Corresponding to the Different Antenna Sizes and Frequencies

Antenna Diameter	10.5 GHz	18 GHz	23 GHz
24 inch	34	38.9	40.5
48 inch	39.9	44.9	46.3
72 inch	43	48.5	49.7

was 11.5 dBm, the signal output at the back of the second antenna, after the transmission over a 10-mile path using two 4-ft antennas, will be

$$\text{Signal output at back of antenna } 2 = 11.5 \text{ dbm} - 52.4 \text{ dB} = -40.9$$

By the time the signal reaches the back of the second antenna, it will encounter another waveguide network, which again will attenuate the signal strength further. Assuming this loss is another 1.5 dB, the signal strength reaching the input stage of the radio (received signal level) is therefore $-40.9 \text{ dBm} - 1.5 \text{ dB} = -40.9 \text{ dBm}$.

Using a hypothetical receiver with threshold sensitivity of -75 dBm at a 10^{-6} bit error rate, the calculated -42.4 dBm received signal level yields a $-47.4 \text{ dBm} - (-75 \text{ dBm}) = 32.6 \text{ dB}$ fade margin. We have a system that works!

By regrouping all the parameters we've touched upon in this section, we can define the parameter flat fade margin (FFM) as

Flat fade margin = Receive signal level − receiver sensitivity

where

Receive signal level (RSL) = Net power gain at transit
point + path attenuation + net
power gain at receiver point

Net power gain at transmit point = Transmitter power output
− waveguide net
+ transmit antenna gain

Net power gain at receive point = Receive antenna gain
− waveguide network loss

When going through this flat fade margin calculation, the reader will notice that there are numerous parameters that control the final receive signal level. Included in the parameters would be the radio, the antennas, as well as the waveguide networks. In summary, we can conclude that

1. an increase in radio transmission power increases receive signal level.
2. a longer waveguide network decreases receive signal level.
3. a larger antenna increases receive signal level.
4. a longer path decreases receive signal level.
5. given the same antenna diameter, a higher-frequency signal gets more antenna gain.
6. higher frequency suffers more signal loss over a given transmission path.

These are some of the parameters we can play with to achieve an acceptable receive signal level. If the user has exhausted all the alternatives, it is time to look into using intermediate repeaters. Depending on different manufacturers, a repeater can be a back-to-back terminal or a regenerator; the difference is that for a regenerator, the radio signal is regenerated at the IF level rather than the

T1 level and is therefore more economical, requiring less circuitry. The system design process is identical to what we have discussed so far, except that now the amount of calculations has just doubled.

There are occasions when the two end terminals are only a couple of miles apart but there is no line-of-sight. In that case, a *passive* repeater alternative is possible. A passive repeater essentially is just two antennas mounted back to back without any radio equipment. The first passive antenna will collect the incoming signal from one direction, feed it through to the second passive antenna using a piece of waveguide, and the second antenna redirects the signal to its final destination. To determine the receive signal level at the destined receiver terminal, we first have to find out the intermediate receive signal level at the passive repeater site. Using the receive signal strength calculation as pointed out before will get us the receive signal level at the output stage of the repeater waveguide network, right before feeding the signal into the second passive antenna. Treat this receive signal level as if it is the output of a transmitter feeding into the second passive antenna, go through the same path lost calculation for the second path, and we'll get the final receive signal level at the final destination. Refer to Figure 4.14 for a graphical explanation.

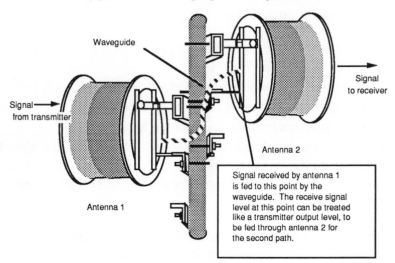

Figure 4.14. Two antennas mounted back to back, configured as a passive repeater.

System Performance

So far, we have discussed how microwave transmission is affected by the atmospheric condition, how the different radio equipment works, how we can put the pieces together to create a system, and even what to do to put a multiple-link network together. But we still don't know how well the overall system performs and how to relate the design with the objective. The rest of this section will tie all the pieces together and answer all the questions.

The most relevant parameter for measuring digital microwave radio performance is the availability rate. As defined in an earlier chapter, availability is the percentage of time a system is performing above a satisfactory rate, such as 99.97 percent over a year at a bit error rate level of 10^{-6}. The unavailability rate of the system is therefore .03 percent.

A system is unavailable for different reasons. From a system design standpoint, we can group them into three categories, each of them responsible for a portion of the outage budget:

1. Multipath fading outage
2. Rain outage
3. Equipment failure

We will address all three outages separately.

Multipath. The system performance of a microwave link can be modeled by the formula

T_O

$$= c \times F_{GHz} \times \frac{1}{4} \times D^3_{mile} \times 10^{-5} \times T_F \times \frac{1}{50} \times 8 \times 10^6 \times 10^{(-CFM/10)}$$

where:

T_0 = one-way annual outage in seconds
c = climate and terrain factor
= 4 over water and Gulf Coast region

$\qquad$ = 1 for average terrain and climate
$\qquad$ = 1/4 for mountains and dry climate
F_{GHz} = microwave frequency in gigahertz
D_{mile} = path distance in miles
$\quad T_F$ = annual average temperature in degrees Fahrenheit
$\qquad$ (Fig. 4.15)
CFM = composite fade margin

The formula basically takes into account the factors that can affect the microwave performance. For example, the climate parameter differentiates the desert from the coastal area (humidity affects microwave bending), the terrain roughness factor differentiates the mountainous region from the flat region (rough terrain allows the wind to stir up the stratified atmosphere), and the path distance parameter heavily penalizes the long path.

The composite fade margin (CFM), however, is not that straightforward to explain. It is almost analogous to the flat fade

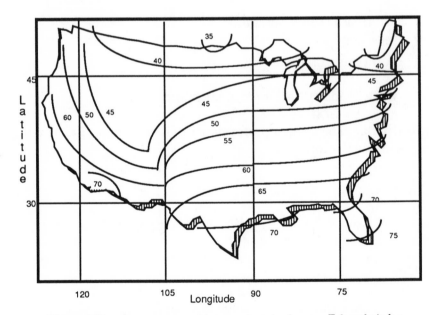

Figure 4.15. Average annual temperature in degrees Fahrenheit for continental United States.

margin we dealt with earlier, where how good a system performs depends on how much margin is allowed to overcome periodic deep fading. Remember that we adopted a flat fade margin of 30 dB for preliminary path design purpose? In an interference-free environment and with a radio equipped with sophisticated equalizers to counter multipath fading, the flat fade margin is very close to the composite margin and is a sufficient starting point.

However, in the real world, there are foreign systems transmitting at the same frequency we want to use, therefore, leading to co-channel interference. Also since most of the 18- and 23-GHz radios use simple modulation schemes such as 4 PSK and are designed for short-haul metropolitan environment, they don't need the sophisticated equalizers to counter multiple fading. Under the circumstances, it is especially important to adopt this composite fade margin parameter for long paths, which accounts for flat fade margin, dispersive fade margin, and interference fade margin. By definition, the composite fade margin is

$$CFM = -10 \log \left[10^{(-FFM/10)} + 10^{(-DFM/10)} + 10^{(-IFM/10)} \right]$$

where

FFM = flat fade margin, or received signal level under normal conditions less receiver threshold sensitivity.

DFM = dispersive fade margin; represents how readily the radio can counter multiple fading conditions.

IFM = interference fade margin, or carrier-to-interference ratio (C/I) - carrier-to-noise ratio (C/N). Frequency coordination typically requires a C/I objective of 65 dB. C/N is equipment specific.

In cases where the dispersive fade margin and the carrier-to-noise ratio are not readily available from the manufacturer, we can conservatively assume the DFM to be 32 dB, which is a typical figure for the more sophisticated 8-PSK system, and a C/N ratio of 25 dB.

In cases where foreign interference is so high that it is not satisfying the C/I objective of 65 dB, we should use the actual C/I figure obtained during the frequency coordination process. Notice a higher interference means lower C/I ratio, thus a lower interference fade margin. We will discuss carrier-to-interference ratio in more detail during the frequency coordination chapter.

For the hypothetical example we have been working with so far, assuming FFM equals 32 dB, DFM equals 32 dB and C/N equals 22 dB, the CFM will be

$$CFM = -10 \, \log \left\{ 10^{(-32/10)} + 10^{(-32/10)} + 10^{[-(65-22)/10]} \right\}$$

or 28.8 dB. Notice the CFM is not that far away from the flat fade margin in this case. However, it is not advisable to substitute the CFM with the flat fade margin directly. As in the case when the flat fade margin is in the 40-dB range, the composite fade margin will be dominated by the dispersive fade margin limitation and remain at the low 30-dB range.

Substituting the 28.8-dB composite fade margin we just derived into the path outage formula, assuming an average climate and terrain factor of 1, a path distance of 10 miles (same distance assumption as the one used to derive the flat fade margin), and an annual average temperature of 50° F, we'll have

$$T_O = 1 \times \frac{18 \, GHz}{4} \times 10^{3}_{mile} \times 10^{-5} \times \frac{50° F}{50} \times 8 \times 10^{6} \times 10^{(-28.8/10)}$$

or 475 sec of one-way outage a year. For a two-way duplex link, this will be doubled to 950 sec a year, or approximately 16 min a year. Since there are 525,600 min a year, the outage is approximately .003 percent of the time, and the availability is 99.997 percent.

There are actual 18-GHz high-capacity systems in Los Angeles, California, spanning from 11 miles to 18 miles, using 6-ft antennas. The outages are observed as predicted, usually during the early morning hours around 2 a.m. The hit duration lasts usually a few seconds and happens approximately once in a week, a perfect example of deep fading at work. On an average, these systems are performing at bit error rates of 10^{-10} and better.

For multiple hops using back-to-back terminals or active regenerator, each hop should be treated independently, with individual outage prediction. The performance of the entire link (end to end) is the sum of the performance of the different hops. If a network link consists of two hops identical to the one we just looked at, the link outage prediction will be doubled to .006 percent.

For a passive repeater, the two hops should be treated as one. The flat fade margin calculation was touched upon in an earlier section. The total path length of the two hops should be used to determine the system availability.

Rain Outage.   Unfortunately, for the 18- and 23-GHz systems used for short-haul network, multipath outage only accounts for a very small portion of the overall unavailability figure. Rainfall outage usually dominates the major portion. Furthermore, rain will stir up the stratified atmosphere, so multipath fading is not possible at the same time; the two events are mutually exclusive of each other.

Modeling of attenuation due to rain on terrestrial links can be very complicated, as research has shown the signal attenuation to be dependent on the rain rate, terminal rain velocity, ambient temperature, and the structure of the raindrop itself. There were experiments and models on raindrop size distributions,[1] index of refraction of water,[2] and so on,[3] and they all have a direct contribution to modeling of microwave rain attenuation. Most of the model data were based on an 18°C temperature, and departure from that temperature can cause the prediction to be off by as much as 20 percent, especially for frequencies below 20 GHz.

It has been observed that a falling rain droplet is not spherical but, rather, spheroidal instead, It is like a spherical droplet being squeezed from the top and the bottom, with a slight ellipsoid shape to it. As a result, a horizontally polarized radio signal suffers more attenuation than does a vertically polarized signal (Fig. 4.16). This difference can be as high as 35 percent. Experiments in

---

[1] J. O. Laws and D. A. Parsons, "The Relation of Raindrop Size to Intensity," *Trans. American Geophysics Union*, vol. 24, 1943, pp. 452–460.

[2] P. S. Ray, "Broadband Complex Refractive Indices of Ice and Water," *Appl. Opt.*, vol. 2, August, 1972, pp. 1836–1844.

[3] Gunn and Kinzer, "The Terminal Velocity of Fall for Water Droplets in Stagnant Air," *J. Meteorol*, vol. 6, 4, 249.

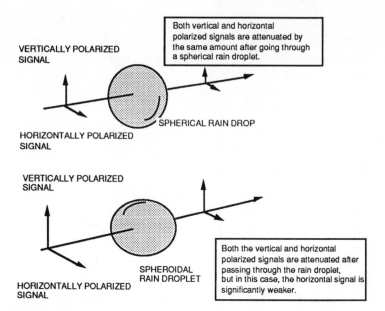

Figure 4.16. A spheroidal raindrop causes signal attenuation to the polarized radio signal.

Europe have established a useful approximation for the relation between attenuation at horizontal and vertical polarizations:

$$A_v = \frac{300 \ A_h}{335 + A_h}$$

where A_v is the attenuation of a vertical polarity signal and A_h is that of a horizontal polarity signal. For example, if a horizontally polarized microwave signal suffers a 10-dB attenuation because of rainfall, a similar signal but only vertically polarized will only suffer an attenuation of

$$A_v = \frac{300 \times 10}{(335 + 10) \text{ dB}}$$

$$= 8.70 \text{ dB}$$

There are models available to predict outage probability taking signal polarity into consideration. However, most of the parameters that the sophisticated models call for, such as raindrop structure, rain intensity based on certain measurement method, and so on, are hard to come by. We need a generic model instead, one that is

Easy to apply.
In good agreement with experimental results in different regions.
Not too critically dependent on the technique used for obtaining the rain data.
Of physical significance.

We will go over a model that satisfies these criteria and appears to yield reasonable agreement with available data.

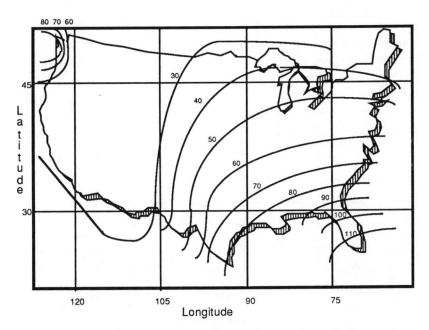

Figure 4.17. U.S. rainfall contours, showing rain intensity exceeded the contour amount .01% of the time. .01% of the time. (International Telegraph and Telephone Consultative Committee, Report 563-2 *Radiometeorological Data*, 1982, p. 114. Washington State contours reflect empirical data by Pacific Bell.)

Step 1. Based on the rainfall distribution map shown in Figure 4.17, determine the location for the radio system. The map shows rainfall intensity along these contours of .01 percent of the time. In other words, rain intensity along these contours is expected to reach that level or more with a probability of .01 percent. For example, in the Boston area, the rain intensity exceeds 42 mm/hr for .01 percent of the time during the year.

Step 2. Experiments have shown that there is a direct correlation between signal attenuation and rain intensity. For practical applications, this relationship can be expressed in terms of attenuation rate a^* (dB/Km) and rain rate R (mm/hr) using the power law

$$a^* = kR^t$$

Based on the assumptions of spherical raindrops, values of k and t have been calculated at a number of frequencies at various temperatures and drop size distributions. Figure 4.18 shows the formula for three different frequencies: 10.5 GHz, 18 GHz, and 23 GHz.[1]

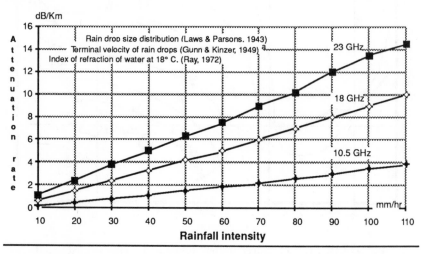

aGunn and Kinzer, "The Terminal Velocity of Fall for Water Droplets in Stagnant Air", *J. Meteorol*, vol. 6, 4, 249.

Figure 4.18. Signal attenuation rate as a result of rainfall, varied by frequency.[2]

[2] In the figure, raindrop size distribution is according to Laws and Parsons (1943), terminal velocity of rain drops to Gunn and Kinzer (1949), and index of refraction of water at 18°C to Kay (1972).

Refer to Figure 4.18 for the specific attenuation rate (a^*) due to rain given a certain rain level and microwave frequency. For example, the 42 mm/hr we have obtained for Boston corresponds to an attenuation figure of 3.8 dB/Km at 18 GHz.

Step 2a. This is a variation of step 2, where the signal polarity as well as the beam azimuth elevation are taken into consideration. This estimate of the attenuation coefficient a^* is certainly more accurate, but it is at the expense of complexity.

Values of k and t for a Laws and Parsons drop size distribution and a drop temperature of 20°C have been calculated by as-

Table 4.3 Parameters for Signal Attenuation Equations
(Signal Polarity Sensitive)

Frequencies	k_h	k_v	t_h	t_v
10.5 GHz	.0119	.0105	1.620	1.247
18.9 GHz	.0649	.0597	1.109	1.078
23 GHz	.1028	.0940	1.075	1.043

suming oblate spheroidal drops—not spherical—aligned with a vertical rotation axis and with dimensions related to the equivalent volume of spherical drops. These values which are appropriate for horizontal and vertical polarizations are presented in Table 4.3 as denoted by k_h, k_v, t_h, and t_v.

Figure 4.19 shows the attenuation rate for the three different frequencies, assuming linear polarizations of vertical and horizontal orientation with no elevation to the transmission beam. For linear and circular polarization, the coefficients k and t can be calculated from the values in Table 4.3 using the equations:

$$k = \frac{k_h + k_v + (k_h - k_v)\cos 2_i \cos 2_j}{2}$$

$$h = \frac{k_h t_h + k_v t_v + (k_h t_h - k_v t_v)\cos 2_i \cos 2_j}{2_k}$$

where i is the path elevation angle and j is the polarization tilt angle relative to the horizontal (j = 45 degrees for circular polarization). Note that for a vertical polarization with no elevation angle,

Vertical Polarization

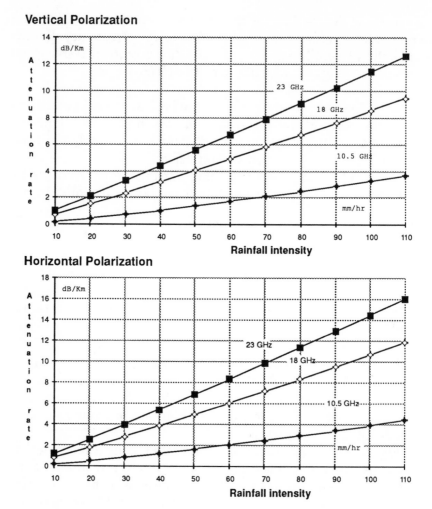

Horizontal Polarization

Figure 4.19. Attenuation rate as determined by rainfall intensity and frequency.

$i = 0$ degreed and $j = 90$ degrees, therefore giving $k = k_v$ and $t = t_v$. For horizontal polarization with no elevation angle, $i = 0$ and $j = 0$, therefore giving $k = k_h$ and $t = t_h$.

In the case where the two end points are located at very different elevations and the transmission path is a short one, the elevation angle i will be a significant one and should be considered in the calculation (Fig. 4.20). In general, the reader may choose to skip step 2a due to its complexity.

Step 3. A path of L_{Km} long will then experience an attenuation

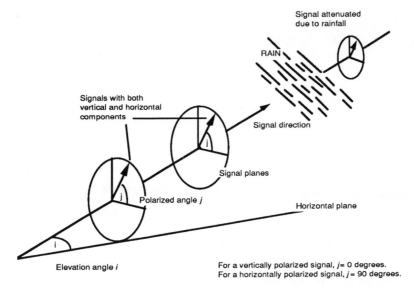

Figure 4.20.   A signal penetrating rain at an elevation angle i, polarized at an angle j.

level of $A_{.01}$ or higher .01 percent of the time, where

$$A_{.01} = a^* \times \frac{L_{\text{Km}} \times 90}{90 + 4\,L_{\text{Km}}}$$

Rearranging,

$$A_{.01} = a^* \times L_{\text{eff}}$$

where

$A_{.01}$ = rainfall attenuation exceeded for .01 percent of the time

a^* = rainfall attenuation rate in decibels per kilometer for rainfall rate exceeded .01 percent of the time

L_{eff} = effective path length to compensate for actual observed result.

= path distance in kilometers $(L_{\text{Km}}) \times 90/(90 + 4L_{\text{Km}})$

Step 4.   If we know the probability of rain attenuation exceeding an $A_{.01}$-dB level is 0.01 percent, we can find out what the attenuation level is for other percentages, maybe through some kind of approximation. Attenuation A_p exceeded for other percentages P is then deduced from the following power law:

$$A_p = A_{.01}\left(\frac{P}{.01}\right)^{-b}$$

$$P = .01 \times \frac{A_{.01}^{(1/b)}}{A_p}$$

where

$$b = .33 \text{ for } 0.001\% \leq P \leq 0.01\%$$
$$= .41 \text{ for } 0.01\% \leq P \leq 0.1\%.$$

If A_p equals the flat fade margin, P will then be the probability that a signal loss due to rain attenuation exceeding the flat fade margin, reaching the outage threshold, therefore, represents the rainfall outage rate. However, note that this approximation *only* applies for the region where unavailability (P) is within the range of .1 percent and .001 percent. In most cases, it is more than sufficient for our purpose.

By inspecting the formula a little more closely, we can redefine the use of the b parameter a little better, as long as the P is still within the applicable percentage region:

$$b = .33 \text{ for } .468 \leq A_{.01}/A_p \leq 1.00$$
$$= .41 \text{ for } 1.00 < A_{.01}/A_p \leq 2.57$$

For our hypothetical example, a^* of 3.8 dB/Km and a 10-mile (16-Km) path gives an $A_{.01}$ of 35.5 dB. Substituting this into the formula for P, where A_p equals the flat fade margin of 32 dB, we have

$$P = .01 \times \frac{35.53^{(1/.41)}}{32}$$

or .0129 percent. In other words, in the Boston area, the expected outage time for this 18-GHz system is .0129 percent. Comparing this figure with the .003 percent figure we calculated for the multiple outage, this rainfall outage certainly dominates the overall outage budget.

Equipment Outage. There are two components controlling the expected equipment outage rate. The first one is the possibility of an equipment outage; the second is the time required to repair the equipment failure.

The possibility of an equipment outage is determined by the reliability of the individual electronic components. In the equipment design process, the manufacturer specifies the kind of components for the equipment, and part of the specification involves the mean time between failure (MTBF) for the components. The individual components mean time between failure rates can then be used to establish an expected mean time between failure for the finished equipment. This figure is then published by the manufacturer. However, be aware that this is only a theoretical figure, and the final performance outcome depends highly on the manufacturing and quality control process, and can be quite different from the theoretical figure.

Furthermore, the mean time between failure of equipment depends on whether it has any redundant feature or not. For hot-standby equipment, that is, every functional part of the equipment is redundant, the mean time between failure rate is a lot higher than the nonhot-standby unit.

Upon equipment failure, the system unavailability clock starts counting. The time necessary to repair the unit depends on:

1. The time required to reach the failed equipment for diagnosis.
2. Equipment failure diagnosis time.
3. Availability of spare parts.
4. Time to replace the failed components.
5. Testing of the circuits before cutting over.

Depending on the user's decision on a maintenance plan, the time required to repair the equipment, called mean time to repair (MTTR) can last from 1 hour to 24 hours, or even days.

To determine the overall unavailability rate, we use the formula

Equipment unavailability

$$= \frac{\text{Mean time to repair}}{\text{Mean time to repair} + \text{Mean time between failure}}$$

For example, if the mean time between failure rate of the equipment is 80,000 hr, and because of well-stocked spare parts and trained technicians, the expected mean time to repair is merely 2 hr. The equipment unavailability rate is therefore only .0025 percent.

Total Performance Budget.   Now that we have all the answers to the questions, we can add up all the expected outage times to get the total expected system outage time.

Total outage rate

= Multipath outage + rain outage + equipment outage

For our example, that is, a one-hop 18-GHz system, 10-mile path in Boston, has an expected outage rate of

Total outage rate (unavailability)

= .003 + .0129 + .0025 = .018 percent

The expected availability rate is therefore 99.982 percent, well within the acceptable design standard.

We have gone through a lot of formulas to reach this point. They are the results of numerous experiments and are excellent models. As a matter of fact, those are the exact formulas used for designing long-haul terrestrial microwave networks; the only differences are in the values of the parameters. To make it easier for the metropolitan network designers, a series of charts and tables are prepared to minimize all the cumbersome calculations. Figure 4.21 is a tree chart that puts all the system performance factors into perspective, while Table 4.4 lists all the relevant equations in one place. All these and the series of charts (Charts 1 to 7), when followed step by step according to instructions in Table 4.5, allow the user to calculate the expected system performance more directly.

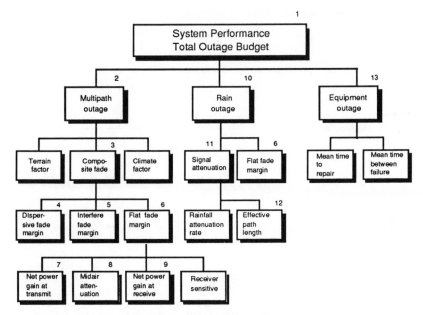

Figure 4.21. A tree chart showing the different parameters as related to overall system performance. (Number next to box represents formula number. Refer to Table 4.5 for the equations.)

Table 4.4 Summary of System Performance Equations

Eq. #	Formulas	
1	Total expected outage	= Multipath outage + rain outage + equipment outage
2	Multipath outage, T_0	$= c \times \dfrac{F\,\text{Hz}}{4 \times D^3_{\text{mile}}} \times \dfrac{10^{-5} \times T_F}{50 \times 8 \times 10^6 \times 10^{(\text{CFM}/10)}}$
3	Composite fade margin (CFM)	$= -10 \log\left[10^{-(\text{FFM}/10)} + \left\{-10^{-(\text{DFM}/10)}\right\} + 10^{-(\text{IFM}/10)}\right]$
4	Dispersive fade margin (DFM)	= Parameter given by the equipment manufacturer
5	Interference fade margin (IFM)	= C/I − C/N at BER threshold
6	Flat fade margin (FFM)	= Receive signal level − receiver sensitivity
		= (Net power gain at transmitter
		+ net power gain at receiver
		+ path attenuation (a negative value))
		− receiver sensitivity
7	Net power gain at transmitter	= Transmitter power output
		− waveguide network loss
		+ transmit antenna gain
8	Path attenuation	$= -[96.6 + 20 \log_{10} (F_{\text{GHz}}) + 20 \log_{10} (D_{\text{miles}})]$
9	Net power gain at receiver	= Receive antenna gain − waveguide network loss
10	Rain outage (P%)	$= .01 \times \left(A_{.01} / A_p\right)^{(1/b)}$, $b = .33$ for $.001\% \le P \le .01\%$;
		$= .41$ for $.01\% \le P \le .1\%$
11	Total signal attenuation due to rain $(A_{.01})$	$= a^* \times L_{\text{eff}}$ (first determine rain intensity for area, then look up rain attenuation rate (Fig. 4.18) for a^*)
12	Effective Path length L_{eff}	$= \dfrac{L_{\text{km}} \times 90}{\left(90 + 4L_{\text{km}}\right)}$
13	Equipment outage	$= \dfrac{\text{Mean time to repair}}{\text{Mean time between failure} + \text{mean time to repair}}$

Table 4.5   System Outage Calculation

Step#	Description	Formula Referenced[1]	Input[2]	Output[3]
1.	Transmitter output power	NA	vendor	____dBm
2.	Transmit frequency	NA	vendor	____GHz
3.	Transmit point waveguide loss	NA	NA	____dB
4.	Transmit antenna gain	NA	vendor	____dB
5.	Net power gain at transmit point	7	1, 3, 4	____dB
6.	Receive antenna gain	NA	vendor	____dB
7.	Receive point waveguide loss	NA	NA	____dB
8.	Net power gain at receive point	9	6, 7	____dB
9.	Receive sensitivity of radio	NA	vendor	____ dB
10.	Path distance	NA	NA	____miles
11.	Midair attenuation (Chart 1)	8	2, 10	____dB
12.	Receive signal level	6	5, 8, 11	____dB
13.	Flat fade margin (FFM)	6	9, 12	____dB
14.	Composite fade margin (Chart 2)	3	13	____dB
15.	Two-way outage due to multipath (Chart 3)	2	2, 10, 14	____%
16.	Rain intensity of the area (Chart 4)	NA	2	____mm/hr
17.	.01% probability that signal will attenuate more this level of $A_{.01}$ (Chart 5)	11, 12	10, 16	____dB
18.	$A_{.01}/A_p$, where A_p = FFM	NA	13, 17	____
19.	Outage due to rainfall (Chart 6)	10	18	____%
20.	Mean time between failure of equipment	NA	vendor	____hr
21.	Mean time to repair the system outage	NA	NA	____hr
22.	Outage due to equipment (Chart 7)	13	20	____%
22.	**Total expected outage**	1	15, 19, 22	____%

[1] Refer to formulas referenced in Table 4.4. The charts are derived based on these formulas.
[2] Refer to the reference step output column. Vendor means vendor specified parameters.
[3] The value according to the corresponding system configuration, formulas, or charts. NA–Not applicable.

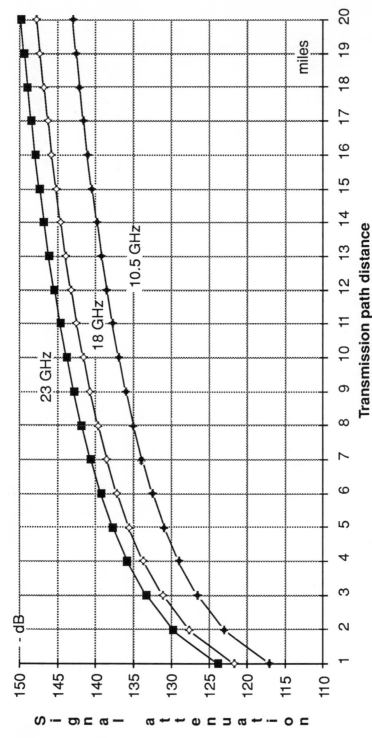

Chart 1. Signal attenuation versus transmission path distance for 10.5-, 18-, and 23-GHz signals.

110

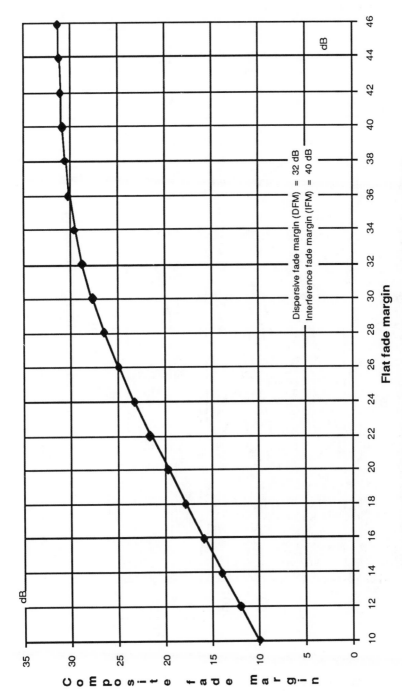

Chart 2. Composite fade margin versus flat fade margin assuming negligible dispersive fade and path interference.

111

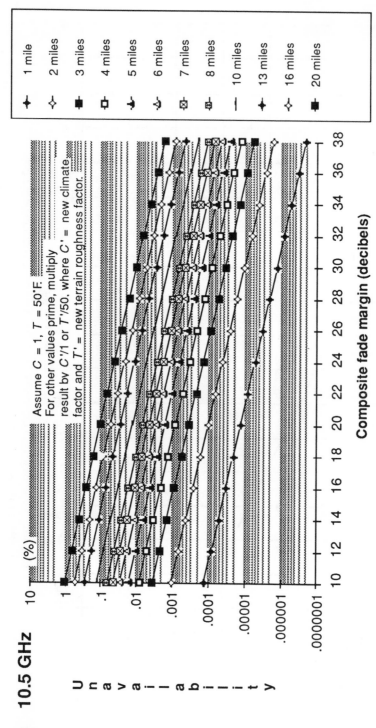

Chart 3a. Unavailability due to path fading (10.5 GHz) versus composite fade margin.

112

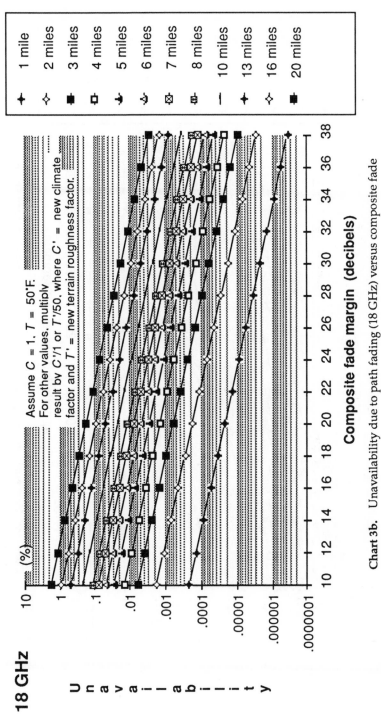

Chart 3b. Unavailability due to path fading (18 GHz) versus composite fade margin.

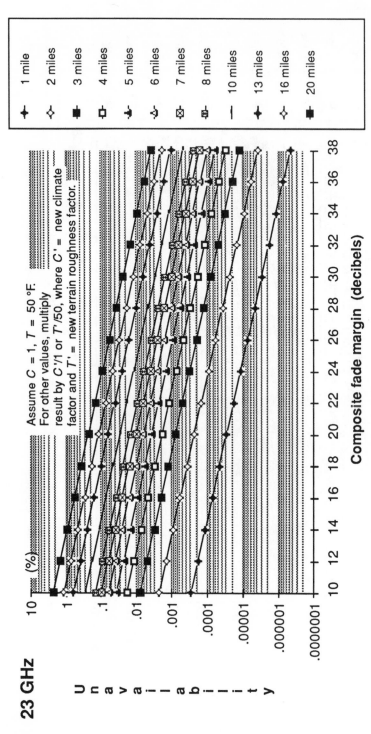

Chart 3c. Unavailability due to path fading (23 GHz) versus composite fade margin.

114

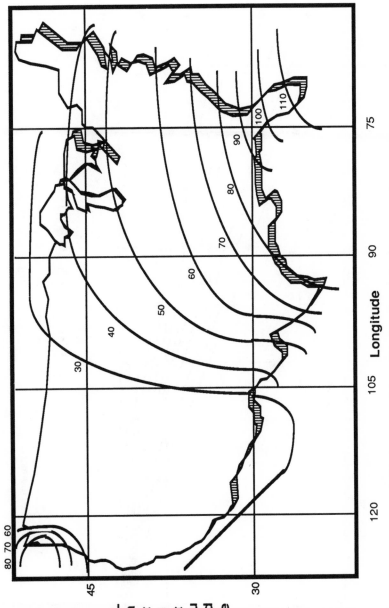

Chart 4. Regional rainfall intensity contour showing areas rain rate exceeded .01 percent of the time.

115

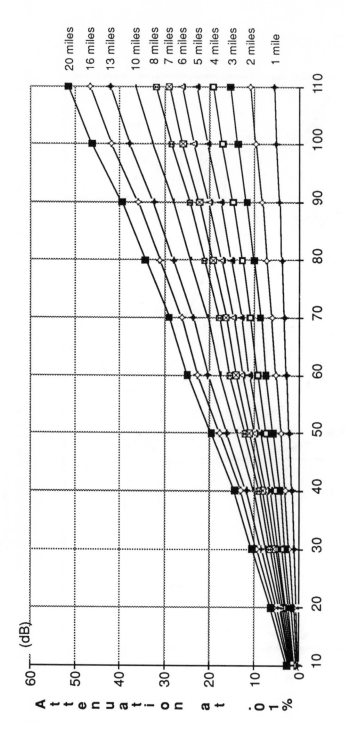

Chart 5a. 10.5-GHz signal attenuation exceeded .01 percent of time versus rainfall intensity.

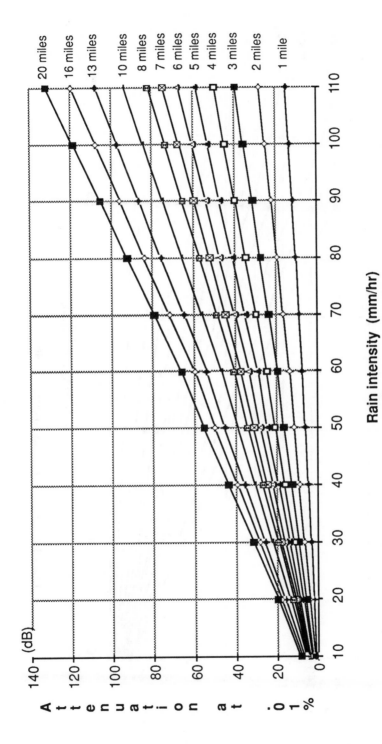

Chart 5b. 18-GHz signal attenuation exceeded .01 percent of time versus rainfall intensity.

117

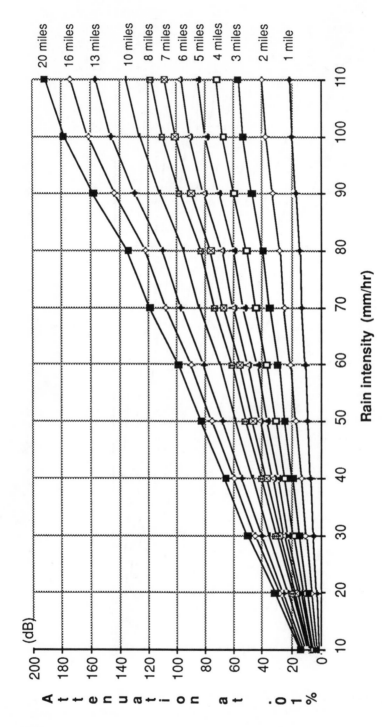

Chart 5c. 23-Ghz signal attenuation exceeded .01 percent of time versus rainfall intensity.

118

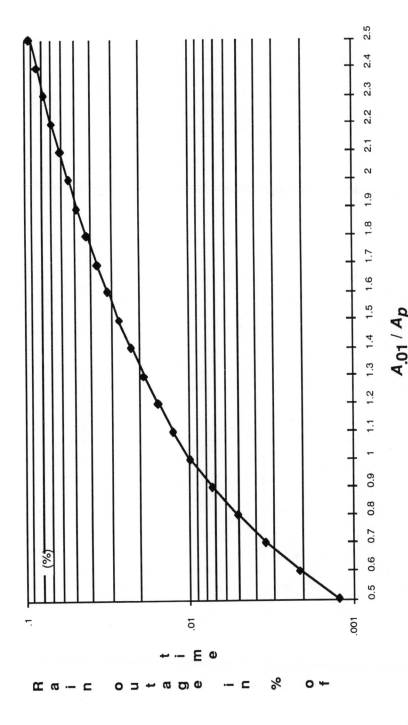

Chart 6. Rain outage in percent of time versus $A_{.01}/A_p$.

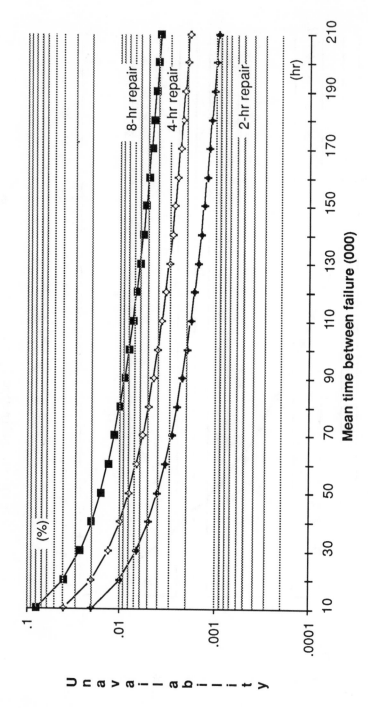

Chart 7. Unavailability versus mean time between failure and mean time to repair.

Example of using the charts.   To go over the charts, let's use the same example as before, that is, using an 18-GHz microwave link over a 10-mile path, with 4-ft antennas on both ends. Referring to Table 4.6 for each step and using the equations as listed in Table 4.4, we have

STEPS

1. Radio output power is 13 dBm according to the equipment specification.

Table 4.6   System Performance Example

Step #	Description	Formula Referenced[1]	Input[2]	Output[3]
1.	Transmitter output power	NA	vendor	13 dBm
2.	Transmit frequency	NA	vendor	18 GHz
3.	Transmit point waveguide loss	NA	NA	1.5 dB
4.	Transmit antenna gain	NA	vendor	44.7 dB
5.	Net power gain at transmit point	7	1, 3, 4	56.2 dB
6.	Receive antenna gain	NA	vendor	44.7 dB
7.	Receive point waveguide loss	NA	NA	1.5 dB
8.	Net power gain at receive point	9	6, 7	43.2 dB
9.	Receive sensitivity of radio	NA	vendor	−75 dB
10.	Path distance	NA	NA	10 miles
11.	Midair attenuation (Chart 1)	8	2, 10	−142 dB
12.	Receive signal level	6	5, 8, 11	−42.6 dB
13.	Flat fade margin (FFM)	6	9, 12	32.4 dB
14.	Composite fade margin (Chart 2)	3	13	28.5 dB
15.	Two-way outage due to multipath (Chart 3)	2	2, 10, 14	.003%
16.	Rain intensity of the area (Chart 4)	NA	2	42 mm/hr
17.	.01% probability that signal will attenuate more this level of $A_{.01}$ (Chart 5)	11, 12	10, 16	33 dB
18.	$A_{.01}/A_p$, where A_p = FFM	NA	13, 17	1.02
19.	Outage due to rainfall (Chart 6)	10	18	.01%
20.	Mean time between failure of equipment	NA	vendor	80,000 hr
21.	Mean time to repair the system outage	NA	NA	2 hr
22.	Outage due to equipment (Chart 7)	13	20	.0025%
22.	**Total expected outage**	1	15, 19, 22	.016%

[1] Refer to formulas referenced in Table 4.4. The charts are derived based on these formulas.
[2] Refer to the reference step output column. Vendor means vendor-specified parameters.
[3] The value according to the corresponding system configuration, formulas, or charts.
NA–Not applicable.

2. Transmit frequency is 18 GHz according to the equipment specification.

3. The waveguide fix loss is dependent on the antenna/radio interface. Assume it to be 1.5 dB for the example.

4. Transmit antenna gain is 44.7 dB for a certain 4-ft antenna, according to the antenna manufacturer specification.

5. Net power gain at transmit point is 56.2 dB, as determined by formula 7 and results from steps 1, 3, and 4.

6. Receive antenna is also 44.7 dB as specified by the manufacturer.

7. Receive point waveguide loss is 1.5 dB, similar to step 3.

8. Net power gain at receive point is 43.2 dB, as determined by formula 9 using results from steps 6 and 7.

9. Receive sensitivity of the radio at 10-6 is 75 dB. This is obtained from the equipment specifications.

10. Path distance has been determined to be 10 miles.

11. According to Chart 1, midair attenuation based on the 18-GHz frequency and the path distance of 10 miles is -142 dB.

12. The signal strength at the radio receiver (RSL) is -42.6 dB, as determined by the power gain at the transmit point and receive point less the power loss over the 10-mile path.

13. The difference, or margin, between the receive signal level and the receiver sensitivity, called the flat fade margin, is 32.4 dB.

14. With the flat fade margin figure of 32.4 dB, determine the composite fade margin based on Chart 2. The assumption, of course, is that the radio has a dispersive fade margin of 32 dB or higher, and there is no interference. The composite fade margin is therefore 28.5 dB.

15. Using the composite fade margin and Chart 3b (18-GHz radio only), the two-way outage probability is determined. Only refer to the line where path distance is 10 miles. If the annual average temperature is not 50°F, the result is adjusted by multiplying the outage probability by T'/50, where T' is the true average temperature. For this example, the average temperature is assumed to be 50°F. Use Chart 3a for 10.5-GHz systems and Chart 3c for 23-GHz systems. For this example, unavailability is .003 percent.

16. To determine the rain outage rate, first determine the rain intensity in the area. For the Boston area, Chart 4 shows that there is .01 percent of chance that the rain intensity will exceed 42 mm/hr.

17. Using Chart 5b (for 18-GHz radio only), we can find that if the rain rate is at 42 mm/hr and the path length is 10 miles, the signal attenuation level is 33 dB. In other words, there is .01 percent of the time when signal attenuation can exceed 33 dB.

18. If there is .01 percent chance that the signal strength can attenuate for more than 33 dB, what is the chance that the signal can attenuate more than the system flat fade margin, therefore causing an outage? First, we have to calculate the parameter A.01/Ap, or 33 dB/32.4 dB = 1.02.

19. Using the ratio obtained in step 18, refer to Chart 6 to determine the rain outage time. For this example, it is .01 percent.

20. Mean time between failure for the radio is 80,000 hr, as specified by the manufacturer.

21. Mean time to repair, based on spares availability and technicians' capability, should average 2 hr. This parameter is customer dependent.

22. From Chart 7, we can determine the outage probability due to equipment failure.

23. The total outage for this link is therefore the sum of steps 15, 19, and 22, totaling .016 percent over a year's period.

Notice the results are slightly different from the calculated results. For example, the rainfall outage percentage is .01 percent instead of the calculated .0129 percent. This is solely a result of visual inaccuracy when reading off a graph. The final result of .016 percent total outage time is approximately 10 percent off the calculated result, a trade-off between time savings and accuracy. Anyway, both tools are available for the designer to adopt, with the graphical approach especially useful for preliminary analysis.

CHAPTER 5
FREQUENCY COORDINATION

Throughout the preliminary system design phase, we have assumed that the presence of foreign interference is negligible to none; therefore, the performance of our system would not be compromised at all. Unfortunately, although this assumption is valid most of the time, the proliferation of the 18- and 23- GHz system, especially in major metropolitan areas such as New York City, Boston, and Los Angeles, has made radio interference an issue.

BACKGROUND ON INTERFERENCE

The Federal Communications Commission (FCC) is the ultimate authority on the use of radio frequencies for any communications use, and one of its roles in the short-haul radio arena is policing the use of the frequencies. It is the federal organization for granting any radio license.

The FCC has assigned the 10.5-, 18-, and 23-GHz spectrums for fixed station microwave communications purpose, in other words, the short-haul microwave system we are dealing with. For each spectrum, it is further divided into transmit and receive blocks, and then channels, each occupying a specific bandwidth. For example, the private 18-GHz spectrum is divided into four transmission blocks:

1. Block A: 17.70 GHz to 18.14 GHz
2. Block B: 18.58 GHz to 18.82 GHz
3. Block C: 18.92 GHz to 19.16 GHz
4. Block D: 19.26 GHZ to 19.70 GHz

Each transmission block is then subdivided into 44 10-MHz channels. For example, the first channel in the A block is 17.705 GHz, pairing with channel 19.265 GHz in the D block for duplex operations. A channel in the B block is paired with another channel in the C block. In total, 88 pairs of frequencies are available. The same kind of grouping applies to the 10.5- and 23-GHz spectrums with a different number of transmit/receive blocks and different channel bandwidths.

Part of the system design process involved picking specific pairs of channels for the radios, with the goal of avoiding frequency interference with a foreign system. If the network consists of only one or two links, the preference is on vertical polarity. However for a complex network, when the user has to maintain a set of maintenance spares at all times, it is better to restrict to a group of frequencies, reusing them if possible to reduce on the sparing requirement; this may lead to potential intrasystem interference.

There are basically three classes of interference: intrasystem interference, external interference, and reflection interference.

Intrasystem Interference

Intrasystem interference arises when a radio within a multihop network interferes with the receiver of a different hop. This is also called overreach interference. Figure 5.1 shows different cases of overreach interference.

In the case when a repeater site uses one frequency to transmit to one direction and uses the same frequency for receiving in a different direction, the antennas will not be able to block all the transmitting signal spilling into the adjacent hop. The solution is of course to use a different pair of frequencies. A better design practice is to divide the transmission block into high band and low band, where each adjacent site will occupy the adjacent band.

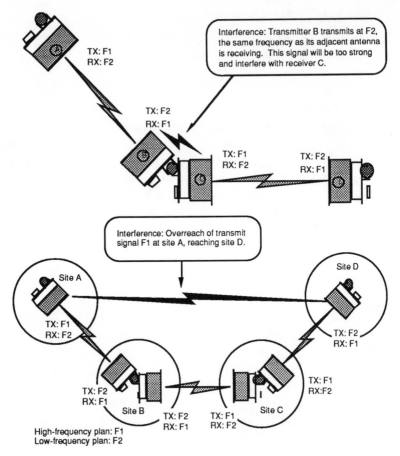

Figure 5.1. Intrasystem interference.

For example, if a network consists of three hops, the first site will choose a transmit frequency in the high band, the second site will choose the transmit frequencies from the low band, and the third site will choose from the high band again. This method is called the high-low plan. Not only does it eliminate signal spilling over to a receiver at the same site, but it also allows the user to grow the network properly, staying with a particular set of frequencies.

Unfortunately, overreach interference still exists for the high-low plan, as shown in the second part of Figure 5.1. The frequency F1, transmitting at site A toward B, is received by site D as well.

During fading condition along path CD, the carrier signal will degrade, thus leading to a decrease carrier into interference margin, and therefore increase in bit errors. The possible solutions are (1) a longer overreach path as compared to the direct CD path, (2) antenna discrimination against the overreach path, and (3) earth or obstacle blocking the overreach path.

External Interference

External interference exists when a foreign system interferes our system. This can happen even if the foreign system is 20 to 30 miles away. Figure 5.2 shows an example of this interference. If we ignore this interference, the worst case will mean our system will never perform at an acceptable bit error rate level.

Calculating the interference level.   Determining the interference level is a very straightforward operation. The calculation is basically similar to calculating a path's received signal level, except that in this case it is the interference path. Furthermore, because the two antennas (interferer and interferee) are not aligned to each other, we account for antenna discrimination as well. According to Figure 5.2, the interference level received at station D, due to foreign station A, is

Interference level at site D = Receive signal level at receive station D, receiving from station A (as if stations A and D antennas are pointing at each other) - antenna discrimination of both antennas

= Transmitter power level of A + antenna gain at A − waveguide loss at A + attenuation over path A to D (negtive value) + antenna system gain at D (assume main beam value) − combined antenna discrimination at both ends

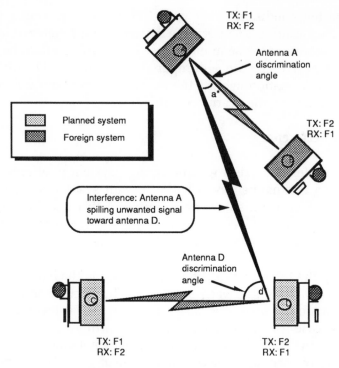

Figure 5.2. Foreign system interference.

The transmit power at station A is basically the radio transmitter output power, less any applicable fix loss (due to the waveguide system), plus the transmit antenna gain. The path loss from station A to D is calculated using the same formula—formula 8 of Table 4.5—in the previous chapter. The antenna system gain at station D, is the antenna gain at D less any waveguide system loss.

Antenna discrimination calculation. The combined antenna discrimination for both ends, the last part of the formula, requires a more detailed explanation. As mentioned in earlier chapters, one of the functions of a parabolic antenna includes the rejection of any incoming foreign signal at an angle to the antenna azimuth. The wider the discrimination angle, the better the noise rejection. Furthermore, if the incoming noise has a different polarization with respect to the incoming signal (cross-polarization), the rejection performance is even better. Figure 5.3 shows a typical antenna envelope pattern published by all antenna manufacturers.

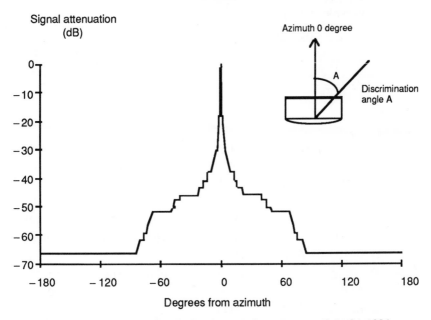

**Gabriel Antenna Model SR4-180
(18 GHz, 48 in.)
HH Pattern**

Figure 5.3.   Antenna discrimination envelope pattern (July 24, 1986 data).

In the example, only an HH pattern is shown, meaning the envelope is only valid for a horizontally polarized antenna receiving a horizontal incoming microwave signal. For example, if a foreign system is interfering at an angle of 10 degrees, the antenna as shown is capable of stepping down the signal (noise) strength by 38 dB. Manufacturers also publish data in the other three polarity combinations, namely, HV, VV, and VH.

Refer back to Figure 5.2. The interference calculation from A to D includes two discrimination angles, the transmit discrimination angle a^* and the receive discrimination angle d^*. Depending on what the polarization combination is between the interferer and the interferee, a combined antenna discrimination factor can be derived using the formulas

Antenna Discrimination (dB)				
Transmit/receive polarity	HH	VV	HV	VH
Transmit discrimination (dB)	A	B	C	D
Receive discrimination (dB)	A'	B'	C'	D'
Combined discrimination (dB)	A + A'	B + B'	A + D' or C + B', whichever is smaller	B + C' or D + A', whichever is smaller

where

1. Regarding the transmitter/receiver polarity, if the transmitter is vertically polarized and the interfered party, the receiver, is horizontally polarized, there is a VH combination.

2. Transmit discrimination (in decibels) is the discrimination factor the transmitter antenna has toward the interferee, as denoted by A, B, C and D.

3. Receive discrimination (in decibels) is the discrimination factor the receiver antenna has toward the interferer, as denoted by A', B', C', and D'.

4. Combined discrimination is the total discrimination factor as determined by the transmit/receive antenna polarity combination and the discrimination angles.

Figure 5.4 shows an example of an interference calculation format typically used by the frequency coordination houses. If the transmit polarity is horizontal and the receive polarity is vertical, the HV combination applies. According to the formula, the total discrimination factor is the smaller of the two values: A + D' (.2 + 60 = 60.2 dB) and C + B' (33 + 60 = 93 dB), yielding 60.2 dB. Applying this to the *net interference level* formula, the interference level is,

INTERFERING PATH BELONGS TO : AAA Communications, Inc.
SITE A: 123 Wilson Tower   TO   SITE B: New World Plaza

Coordinates: 34 03 41 Longitude   34 02 54 Longitude
               118 18 10 Latitude   118 15 27 Latitude

Distance = 2.75 miles	Power = 23.0 dBm *(a)*
Bearing = 108.91 degrees	Antenna gain = 38.9 dB *(b)*
Loading = 672 channels dig.	Waveguide loss = 0 dB *(c)*
Stability = .003%	
Ground = 220 ft AMSL	Centerline = 306 ft

Transmit frequencies at site A : 19320.00 H 19600.00 V

DISTURBED PATH BELONGS TO : XYZ Industries
SITE C: 234 Broadway Ave   TO   SITE D: Kings Point

Coordinates: 34 02 53 Longitude   34 02 51 Longitude
               118 15 14 Latitude   118 15 17 Latitude

Distance = .07 miles	Power = 0.0 dBm
Bearing = 235.53 degrees	Antenna gain = 38.7 dB *(d)*
Loading = 672 channels dig.	Waveguide loss = 0 dB *(e)*
Stability = .003%	Normal RCV LVL = –32.2 dB *(f)*
Ground = 260 ft AMSL	Centerline = 170 ft

Transmit frequencies at site C : 19320.00 V 19600.00 H

Interfering path length (from A to D)	2.92 miles
Interfering path bearing	108.96 degrees
Interfering path attenuation	–131.31 dB *(g)*

Antenna information/discrimination	GAIN	VV	VH	HH	HV	
Site A: Mark antenna HP-17024WD	38.9	0.2	33	0.2	33	
Site D: Cablewave DAX2-190	38.7 *(f)*	60	60	60	62	
Total antenna discrimination *(h)*		60.2	62.2	60.2	60.2	
Net interference + level *(a + b – c + g + d – e – h)*		–90.9	–92.9	–90.9	–90.9	
						OBJ
C/I ratio [*(f – (a + b – c + g + d – e – h))*]		58.7	60.7	58.7	58.7	65

Figure 5.4.   A frequency interference calculation. (Formulas shown
in italics.)

therefore,

Interference level = Transmitter power level of A
+ antenna gain at A
− waveguide loss at A
+ attenuation over path A to
   D (negative value)
+ antenna system gain at D
   (assume main beam value)
− combined antenna discrimination
   at both ends

= 23 dB + 38.9 dB − 0 dB
   + (-131.31dB) + 38.7dB − 0 dB
   − 60.2 dB

= 90.91dB.

The carrier-to-interference ratio.   Notice Figure 5.4 goes beyond calculating an interference level figure. It introduces two new parameters called carrier-to-interference ratio (C/I) and a C/I objective. From a system design viewpoint, the absolute interference level alone does not determine a system's performance level. We have to compare it with the strength of the incoming intended signal. The parameter carrier-to-interference ratio shows how serious this interference is in relation to the carrier signal:

C/I = Receive signal level at receiver − interference level at same receiver

For example, if the normal receive signal at a receiver is −32.2 dB, and the interference level is −90.9 dB, the C/I ratio is then −32.2 − (−90.9) dB, or 58.7 dB (as shown in Fig. 5.4). For frequency coordination purposes, this 58.7-dB ratio is deemed inadequate; an objective of 65-dB is often sought by the manufacturer. If we want to determine what the impact is on system performance as a result of not meeting the 65 dB objective, we can go back to the interference fade margin (used in the composite fade margin) parameter discussed in Chapter 4 and substitute the 65-dB figure with the actual C/I figure. Depending on the network objective, the designer may find that, even though the C/I objective is missed by as much as 10 dB, the performance level may still be acceptable.

Now if we step back and look at the overall objective, we'll notice the key to an interference-free environment is not to reduce the absolute interference level but to increase the overall C/I performance. This can be done through (referring to Fig. 5.2 again):

1. Inncrease transmission power at C (by increasing antenna size and gain); and therefore increase the receive signal strength reaching antenna D.
2. Look into an alternate antenna for site A to achieve a better signal discrimination at the angle a^*.
3. Select an alternate site to achieve a longer interference path length.
4. Select an alternate site to achieve a larger discrimination angle for a^* or d^*.
5. Select an alternate site to avoid a line-of-sight between sites A and D.

Reflection Interference

In a metropolitan environment, the numerous building structures can become unwanted reflection surfaces to any microwave system. A foreign signal may be reflected into the receiver at a relatively high level and overwhelm the intended incoming information. Furthermore, multiple buildings along the signal path may cause multiple reflections of a portion of the original signal, causing a significant delay to become noise instead. Figure 5.5 shows an example of building reflection.

Since how a microwave beam reflects off a surface depends highly on the kind of reflection surface, the angle of incidence, and so on, it is very difficult to predict on paper how severe (or negligible) a reflection interference can be. Furthermore, the ever-changing urban skyline can make any paper prediction or even on-site signal measurement effort fruitless. The only logical and economical alternative a designer has is build and hope: that is, building the system and hoping there is no problem with any reflection. Though it may sound unscientific and irresponsible, consider the fact that the chance of having a reflection problem is as remote (or as possible) as having a brand-new skyscraper being built right in the middle of the signal path. It is a risk worth taking.

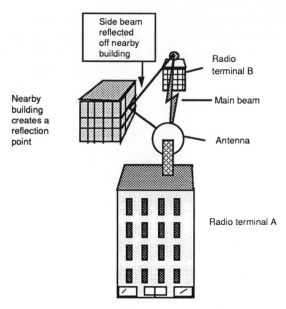

Figure 5.5.   A building at the middle of the path reflects part of the signal, thereby causing interference.

FREQUENCY COORDINATION PROCEDURES

After learning about the frequency spectrum, the importance of minimizing system interference, the user now confronts the ultimate question: How does one go about coordinating all the frequency-related activities to find out where these foreign systems are? And better yet, how does one prevent future foreign systems from interfering with the current ones?   Imagine all the record-keeping efforts involved. It is precisely these problems that prompted the FCC to establish a frequency coordination guideline.

Background on Frequency Coordination

Point-to-point microwave systems have been used to provide common carrier communications since the late 1940s. Until 1971 when the prior coordination requirement was established, microwave applications were filed with the FCC, either without any prior frequency coordination or, at best, with some informal coordination with selected parties. Applications accepted for filing by the FCC were listed in its FCC Public Notices, published weekly. When other microwave users reviewed the Public Notices and noted that a particular application might result in objectionable interference to their facilities, coordination studies were conducted. Often, the studies could not be completed within the usual 30-day Public Notice period, and parties filed motions with the FCC to delay commission action on those applications. If the studies subsequently showed interference not to be a problem, the FCC was so notified. If, on the other hand, potential interference was found to be objectionable, protests were filed against the application(s) in question. Obviously, this "coordination" process proved cumbersome, both for the microwave users and the FCC.

During the 1960s AT&T and Western Union, by virtue of their extensive development of microwave radio networks, assumed dominant positions with respect to frequency coordination. Both companies realized that effective radio system planning depends on some form of frequency coordination and accurate information on other companies' radio facilities. AT&T and Western Union each created a catalog of radio stations which listed all pertinent frequency coordination data. By exchanging these system catalogs, AT&T and Western Union were able more effectively to select "interference-free" frequencies before filing FCC applications for use of those frequencies. To maintain the accuracy of their frequency coordination databases, these companies reached informal agreements to coordinate new proposals before their implementation and to exchange data periodically on each others' microwave network. Since these organizations represented the most prominent users of microwave systems, their pooling of coordination data greatly facilitated the planning and engineering of new systems.

During this period, other smaller companies attempted frequency coordination using published listings of licensed frequencies or the existing FCC records, which at that time consisted of copies of licenses, radio maps, and other files. The effectiveness of this method was limited by the accuracy of the files as well as the large amount of manual researches it required. Many applications coordinated in this manner were the subject of protests of other microwave users who claimed interference to their facilities.

During the 1969 to 1971 time frame, several consulting firms offering frequency coordination services were formed. Many of the smaller users hired these consultants as their frequency coordination agents and this helped reduce the "hit or miss" nature of frequency planning. At the same period, the start-up of special common carriers like MCI and US Sprint resulted in a dramatic increase in the number of construction permit applications filed. The FCC quickly realized that the process of frequency coordination, which then involved many postapplication legal petitions and FCC hearings, had to be improved. Thus the FCC made prior frequency coordination of point-to-point microwave radio applications a requirement effective July 15, 1971. Since that time, prior coordination requirements have been extended to other communications services, including the private operational fixed microwave service, to which the short-haul metropolitan microwave network belongs.

Frequency Coordination Responsibility

Basically, the responsibility of a new system user during the frequency coordination process consists of the following:

1. *Interference Analysis.* Perform interference analysis against all existing systems using the same frequency spectrum, within an area of 150 miles of the proposed system. Identify any potential interference case and redesign the proposed system to eliminate them.

2. *Prior coordination notification (PCN).* Notify all users in the service area regarding the proposed system. The prior coordination notification distributed to other parties should contain sufficient technical information to allow them to assess accurately the potential for interference. Generally, the information includes station location, channel frequencies and polarizations, type of emission, transmit power, antenna characteristics, and antenna height (centerline). The existing system users have up to 30 days to respond if there is any objection. The new system proponent basically has to answer all the objections satisfactorily, supplying any additional information if necessary, to clear the case. Figure 5.6 shows an example of the public notice.

3. *Subsequent notification and response.* If potential interference problems are reported to the initiating coordinator, he or she

PCN Date: 4/3/1988   MICROWAVE PATH DATA SHEET

Station	(Los Angeles Ca)	(Anaheim Ca)
Path status	Proposed for prior coordination	
Call sign	WLC6500	WLC6501
Latitude (D-M-S)	34 01 13.0	34 07 09.0
Longitude (D-M-S)	118 25 23.0	118 23 30.0
Ground elevation (ft.)	117	1350
Path azimuth (degree)	14.79	194.81
Path distance (mile/Km)	7.05 miles/11.35Km	
Antenna		
Primary transmit	Digital Microwave	Digital Microwave
	086-423218	086-423218
FCC code	DD0180	DD0180
Antenna gain (dB)	38.5	38.5
Center line (ft. AGL)	190	40
Diversity	Same as transmit	
FCC code		
Antenna gain (dB)		
Center line (ft. AGL)		
Equipment		
Name	Digital Microwave	Digital Microwave
	Corp.	Corp.
Model	DYH6RMDMC23-01	DYH6RMDMC23-01
FCC code	2YAM02	2YAM02
Emission	25000A9Y	25000A9Y
Loading	96 channel, digital	96 channel, digital
Stability (%)	0.030	0.030
Power (dBm)	15	15
Receive level (dBm)	– 48.5	– 48.5
Fixed losses (dB)	1.5	1.5
Free space loss (dB)	140.5	
Transmit frequencies	22475.00V 22525.00G	21275.00V 21225.00G

Figure 5.6. The prior coordination notice before filing with the FCC.

may be able to modify the technical characteristics of the proposed system to resolve those problems. The modifications,
however, may create new problems for some other parties
who formerly had no objection to the proposal. Therefore,
when changes are made in coordination data, additional notifications should be distributed to all the parties originally
notified. In some cases the changes may be minor and may
have a seemingly negligible impact on some or all of the other

parties. However, when any changes are made involving coordination data, notifications should be distributed to give other parties the opportunity to study the changes and to allow them to maintain accurate data on existing and proposed systems for future coordination purposes. Where subsequent coordination notifications involve relatively minor changes, responses should generally be made within 30 days.

4. *FCC filing.* A radio license can be obtained through the FCC by submitting Form 402 (Appendix). When filing with the FCC, the applicant must certify that prior frequency coordination has been completed. The application must also include the supplemental showing, with a list of parties (licensees and other applicants) with whom coordination was conducted, clearly identifying those parties to whom notification was made. The supplemental showing is usually ready within 4 days after the path is studied by the frequency house and can be filed with the Form 402 immediately. Figure 5.7 shows one such supplemental showing.

   If an interference problem emerges during the FCC filing, the applicant must clear the objection immediately or risk the rejection of the application. Any major corrections such as changing equipment or frequency will require an entire new filing.

   The lead time for a license is usually 60 to 90 days after submission. Once the FCC has granted the license (but not before), the applicant can start installing the system. However, any major changes made to the system during the construction phase require immediate amendment of the application, and that means further delay in the license granting process. The applicant may not activate the system unless it conforms exactly to what's specified in the license.

5. *Granting of FCC license.* Once the license is granted, the proponent has up to 12 months from the date of grant to construct the system, or the authorization expires automatically. Upon construction, the proponent must inform the FCC district engineer 5 days before the system is ready for power up and testing. A copy of the license must be posted at the corresponding transmitter site.

FREQUENCY COORDINATION, INC.
1240 Pennsylvania Ave
Reston, VA 22092
(703)222-2222

SUPPLEMENTAL SHOWING, PART 94(B)

Advance Technology International
Los Angeles–Anaheim, California, 23 GHz Terrestrial Microwave

Pursuant to Part 94 of the FCC Rules and Regulations, the above-referenced microwave route was coordinated with the existing licensees and applicants whose facilities could be affected. Coordination data were forwarded on April 3, 1988.

The following parties or their designated coordination agents were notified:

> Spectrum Planning Incorporated
> Comsearch Incorporated
> University of Southern California
> Shultz Steel Company
> Vector General, Inc.
> San Bernardino County, CA

There are no unresolved interference objections with the stations contained in these applications.

Respectfully submitted
Frequency Coordination, Inc.

John Doe
Staff Engineer

Figure 5.7. The supplemental showing accompanying the FCC filing.

In summary, it is critical for the user to be totally familiar with the frequency coordination process, as any mistakes made can cost up to 30 days of delays on top of additional project costs. It is highly advisable that first time (or small-scale) users hire the frequency coordination companies, as they are specialized in performing interference analysis, coordinating frequencies, filing FCC applications, as well as protecting the proponent's frequencies from interference due to future systems. The key issue is that the new user has to know what to look for when talking to these coordination companies.

THE APPLICATIONS

As mentioned, the frequency coordination process involves sending out the prior coordination notice (PCN) as well as FCC Form 402. This section will briefly go over the two forms formats. The FCC instructions for filing Form 402 can be found in the Appendix.

There is no specific rule about the format on the prior coordination notice. However, the format shown in Figure 5.6 is a generally accepted standard throughout the industry. Some of the important information on the PCN are

1. *Date*. The date the PCN was issued. Any system proposed after that date may interfere with the orginal system. Potential interference must be rectified by the later applicants at their own cost.

2. *Station names*. Station name is the name of the terminal where the radio is located. It can be names such as "headquarters" or "New York City 1."

3. *Call sign*. This is the station identification number issued by the FCC, very similar to the ones used by the broadcasting station such as WABC. In the case of private terrestrial microwave users, the call sign always has three alphabets followed by four digits. For common carriers, the call sign would have three alphabets followed by three digits instead. Each call sign can only associate with one physical location, and vice versa, under the same station owner. When filing a new system, the call sign information will be left blank because the FCC has not approved and assigned a call sign yet. However, if a network is being expanded, adding new radio channels to an existing station, the call sign granted by the FCC for the existing system should be used here, thus notifying the FCC that the proposed system is going into an approved operating facility.

4. *Longitude and latitude.* Numbers expressed in units of degree, minute, and second to represent the exact location of the station. The figures can be read off topographical maps. This set of numbers is used to coordinate with other foreign systems, and their accuracy cannot be overlooked. A difference of one second may mean a difference of almost 100 ft, or even a difference of several degrees in antenna azimuth.

5. *Ground elevation.* This can be obtained from a topographical map.

6. *Path azimuth and path distance.* The path azimuth indicates the orientation of the antenna measured in degrees from true north. Path distance is the distance between the two stations. The frequency coordination company usually calculates these two figures based on the longitude and latitude numbers supplied by the customer. The calculated figures should always be checked against the measured ones taken off the topographical map. This serves as a check on the accuracy of the longitude and latitude numbers.

7. *Antenna information.* This can be readily obtained from the radio/antenna manufacturer. A lot of times, the frequency house has all that information on file. The important thing is to make sure the proper antenna model and size are used. The center line of the antenna represents the height of the antenna center feed horn above the ground level. In the case of a 4-ft antenna mounted with its bottom 10 ft above a building roof level, the center line will be the building height plus the 12 ft, 10 ft from the roof top to the antenna base, and another 2 ft to the center feedhorn.

8. *Primary antennas and diversity antenna.* This is the information on antennas the system is using for the path. Diversity antenna is usually used on long path (20 miles or further) to minimize multipath fading phenomena and is almost never used in the metropolitan network environment. As a matter of fact, the majority of the 18- and 23-GHz radios are not designed to handle diversity antenna.

9. *Equipment information.* All radio equipment must be the type accepted by the FCC. Upon acceptance, the FCC will issue an identification code for the equipment. Parameters such as emission designator, channel loading, frequency stability and

output power are all unique to that specific radio. This information can be obtained from the radio manufacturer directly and is usually on file with the frequency coordinating companies.

10. *Received level, fixed loss, and free space loss.* These parameters were discussed in the previous chapter on path design. The figures are also calculated automatically by the frequency house during the PCN phase.

11. *Transmit frequencies.* This indicates the exact frequency and polarity the radio will be transmitting at the station. For example, a transmit frequency of 22475.00 V at site A and 21275.00 V at site B means site A will be transmitting at 22.47500 GHz, vertically polarized and receiving a signal of 21.27500 GHz from site B, also vertically polarized. On occasions where there is a potential need for additional radio channels, the applicant may "reserve" this during the PCN phase. This is accomplished by using the two alphabets "G" and "U" in place of "H" and "V" after the frequency number: "G" stands for a growth channel, horizontally polarized; "U" stands for a vertically polarized growth channel. Occasionally, the symbol "B" will be used, representing a growth frequency of both polarities desired for the future.

Once the PCN is prepared, a supplemental showing can be generated within 14 days by a frequency house. The Form 402 filing can then proceed immediately using the supplemental showing. Most of the information in the PCN notice has to be included in Form 402; also required is information on the applicant's background. Part of the Appendix goes over the 402 Form filing instructions in more detail.

The difference between the PCN format and Form 402 format is not a trivial one. While the PCN form gives the information on one path, between two separate locations, the 402 Form gives the information on one location with multiple signal paths transmitting from that one site. For instance, if the network consists of three locations, A, B, and C, where A is linked to B and B is linked to C, the PCN will show two paths, one from site A to site B and another one from site B to site C. Form 402, however, will involve three different forms, the first two applications cover location A and C, respectively, both showing they are transmitting toward location B, and the third 402 Form will cover location B, with radios transmitting toward sites A and C.

Another difference between the PCN and the 402 Form is in the frequency area. While the PCN provides a mean to reserve future growth frequencies, the Form 402 does not. Any frequency shown on Form 402 must be activated within 12 months of granting the application.

Figure 5.8 shows a completed Form 402 application, reflecting the PCN path example shown earlier.

UNITED STATES OF AMERICA FEDERAL COMMUNICATIONS COMMISSION
GETTYSBURG, PA 17326
APPLICATION FOR STATION AUTHORIZATION IN THE
PRIVATE OPERATIONAL FIXED MICROWAVE RADIO SERVICE

Approved by OMB
3060-0064
Expires 10/31/89

FOR COMMISSION USE ONLY		
FILE NUMBER:	NEPA: ☐ MINOR ☐ MAJOR Date of NEPA Public Notice	SEND TO ASB: ☐ YES ☐ NO
PUBLICATION NOTICE REQUIRED: ☐ YES ☐ NO		ASB REMARKS:

FOR APPLICANT: Use FCC Form 402 Instructions dated August, 1985 or later for reference in completing form.

SECTION I—IDENTIFICATION INFORMATION

1. NAME OF APPLICANT:

3. CALL SIGN: (If application refers to an existing station)

2. MAILING ADDRESS: (No., street, city, state, ZIP code)

4. LICENSEE IDENTIFICATION NUMBER: (If previously assigned by the Commission)

5A. NAME OF PERSON TO CONTACT REGARDING APPLICATION:

☐ Check here if you are a current licensee and your mailing address, item 2, IS NOT the address on file.

5B. TELEPHONE NUMBER OF THE CONTACT:

6. TYPE OF APPLICANT: ☐ INDIVIDUAL   ☐ ASSOCIATION   ☐ PARTNERSHIP   ☐ CORPORATION   ☐ GOVERNMENTAL ENTITY

7. CLASS OF STATION: (enter code from instruction 7)

8. RULE SECTION UNDER WHICH YOU ARE ELIGIBLE:

9A. PURPOSE OF APPLICATION:

(A) ☐ NEW STATION   (B) ☐ MODIFICATION (SEE 9B & 9C)   (C) ☐ MODIFICATION WITH RENEWAL (SEE 9B & 9C)   (D) ☐ ASSIGNMENT OF AUTHORIZATION   (E) ☐ OTHER (SPECIFY) ▶

9B	PATH	ACTION			OLD VALUE OF KEY ITEMS CHANGED			
	A	☐ ADD	☐ CHANGE	☐ DELETE	20	30	31	32
	B	☐ ADD	☐ CHANGE	☐ DELETE	20	30	31	32
	C	☐ ADD	☐ CHANGE	☐ DELETE	20	30	31	32
	D	☐ ADD	☐ CHANGE	☐ DELETE	20	30	31	32

9C. DESCRIBE ANY OTHER CHANGES:

10. Will the use of this station be shared by another Party?
☐ YES   ☐ NO

SECTION II—ANTENNA INFORMATION

11. LOCATION OF TRANSMITTING ANTENNA STRUCTURE:

A. NUMBER AND STREET: (or other specific indication)

B. CITY:	C. COUNTY:	D. STATE:

E. COORDINATES:

LATITUDE: (Degrees, Minutes, Seconds) NORTH	LONGITUDE: (Degrees, Minutes, Seconds) WEST

12A. Is the antenna to be mounted on an existing antenna structure? If yes, answer items 12B, C, D, and E.
☐ YES   ☐ NO

12B. Will the antenna increase the height of the existing structure? If yes, by how many feet?
☐ YES   ☐ NO   FT.

12C. NAME OF CURRENT LICENSEE USING STRUCTURE:

12D. CURRENT LICENSEE'S RADIO SERVICE:	12E. CURRENT LICENSEE'S CALL SIGN:

13. For antenna towers or (poles) mounted on the ground:

Enter the overall height above ground of the entire antenna (or pole) including all antennas, dishes, lightning rods, obstruction lighting, etc., mounted on it .... FT.

14. For antennas or antenna towers (or poles) mounted on a supporting structure such as a building, water tower, smoke stack, etc:

A. What is the overall height above ground of this supporting structure? Include in this height any elevator shafts, penthouses, lightning rods, lights, etc., which are not a part of the antenna tower (or pole) .... FT.

B. How many feet does the antenna tower (or pole) (including all antennas, dishes, lightning rods, lights, etc.) increase the height of the supporting structure in Item 14A? If this antenna or antenna tower (or pole) does not increase the height of the supporting structure, enter zero (0) .... FT.

C. What is the overall height above ground of this supporting structure plus the antenna tower (or pole)? .... FT.

15. Give the ground elevation above mean sea level at the antenna site .... FT.

16A. NAME OF NEAREST AIRCRAFT LANDING AREA:	16B. DIRECTION AND DISTANCE TO NEAREST RUNWAY:

FCC 402
March 1987

Figure 5.8.

17A. Has notice of construction been filed with the FAA on FAA Form 7460-1? If yes, answer items 17B, C, and D.   ☐ YES   ☐ NO

17B. NAME UNDER WHICH YOU FILED:	17C. FAA REGIONAL OFFICE: (City)	17D. DATE FILED:

18. Would a Commission grant of your application be an action which may have a significant environmental effect as defined by Section 1.1307 of the Commission's Rules? See Instruction 18. If you answer yes, submit the statement as required by Sections 1.1308 and 1.1311.   ☐ YES   ☐ NO

19. If this is an existing station, enter the year it was first licensed:

SECTION III—TECHNICAL INFORMATION

NAME OF ITEM	A	B	C	D
20. Frequency (MHz)				
21. Bandwidth (kHz) and Emission Type				
22. Type of Message Service				
23. Initial Baseband Channel Loading				
24. 10 yr. Projected Baseband Channel Loading				

TRANSMITTER INFORMATION

NAME OF ITEM	A	B	C	D
25. Transmitter Operating Frequency				
26. Antenna Gain (dBi)				
27. Effective Isotropic Radiated Power (dBm)				
28. Beamwidth (Degrees)				
29. Height to Center of Final Radiating Element (Ft)				
30. Polarization				
31. Azimuth to Receive Site or Passive Repeater (PR) No. 1 (Degrees)				

RECEIVE SITE INFORMATION

NAME OF ITEM	A	B	C	D
32. Receiving Station's Call Sign				
33. Receiving Antenna Gain (dBi)				
34. Median Received Signal Level at Input to the Receiver (dBm)				
35. Latitude N (Degrees, Minutes, Seconds)				
36. Longitude W (Degrees, Minutes, Seconds)				
37. Ground Elevation AMSL (Ft)				
38. Height to Center of Receiving Antenna (Ft)				

PASSIVE REPEATER NO. 1 INFORMATION (IF ANY)

☐ If you have two or more passive repeaters on the same transmission path, check this box and answer items 39-46 on an additional FCC Form 402 or a separate sheet of paper for the second and successive passive repeaters.

NAME OF ITEM	A	B	C	D
39. Latitude N (Degrees, Minutes, Seconds)				
40. Longitude W (Degrees, Minutes, Seconds)				
41. Ground Elevation AMSL (Ft)				
42. Overall Height of PR Structure Above Ground (Ft)				
43. Dimensions (Ft X Ft) or Beamwidth (for dishes) (Degrees)				
44. Height Above Ground to Center of PR (Ft)				
45. Polarization				
46. Azimuth to Receive Site or Next PR (Degrees)				

SECTION IV—CERTIFICATION

1. Applicant waives any claim to the use of any particular frequency regardless of prior use by licensee or otherwise.
2. Applicant will have unlimited access to the radio equipment and will control access and exclude unauthorized persons.
3. Neither applicant nor any member thereof is a foreign government or representative thereof.
4. Applicant will utilize type accepted radio equipment and antennas of correct specifications.
5. Applicant certifies that all statements made in this application and attachments are true and complete.

WILLFUL FALSE STATEMENTS MADE ON THIS FORM OR ATTACHMENTS ARE PUNISHABLE BY FINE AND IMPRISONMENT U.S. CODE TITLE 18 SECTION 1001

TYPED NAME	TITLE

SIGNATURE of individual, partner, official of a governmental entity, officer or authorized employee of a corporation, or officer who is also a member of the association	DATE

FEES: Except for Public Safety and Governmental applicants, each microwave station application must be accompanied by a $135 fee and filed at: FCC, Microwave Service, P.O. Box 360850M, Pittsburgh, PA, 15251-6850. For information on fees call (717) 337-1212.

NOTICE TO INDIVIDUALS REQUIRED BY PRIVACY ACT OF 1974 AND THE PAPERWORK REDUCTION ACT OF 1980
Sections 301, 303, and 308 of the Communications Act of 1934, as amended, (licensing powers) authorize the FCC to request the information on this application. The purpose of the information is to determine your eligibility for a license. The information will be used by FCC staff to evaluate the application, to determine station location, to provide information for enforcement and rule-making proceedings and to maintain a current inventory of licensees. No license can be granted unless all information requested is provided. Your response is required to obtain this authorization.

FCC 402
March 1987

Figure 5.8.

CHAPTER 6
FIELD SURVEY

So far, we have dealt with design work that is done in an office, with assumptions on line-of-sight, certain antenna size, and so on. However, this design phase is not complete unless a field survey is performed. Its importance cannot be emphasized enough, as numerous mistakes are made simply because no site survey was done to verify the assumptions, or the surveys were done incorrectly. Remember, by the time the user finds out, it may very well be too late to pull back because all the equipment has been delivered and installers are on site ready to do the installation work.

Site survey is more than looking out an office window to see if the far-end terminal is in sight. A lot of mistakes can be made because of that simple look. For example, it is possible for the roof to have so much mechanical equipment that it is impossible to mount the antenna securely at that part of the building, and an alternate location on the roof means trees are in the way, blocking the line-of-sight. Another potential problem may be that to obtain a line-of-sight, the antenna has to be placed at the far end of the roof, causing the total cable length from the radio to the indoor modem to exceed the radio's cable distance limitation.

Obtaining line-of-sight for a microwave radio means very different things to different people. As we have discussed in earlier chapters, line-of-sight means clearing any obstacles by a certain margin, taking into consideration the Fresnel zone and K factor clearances. The key is to use a surveyor who knows what to look for and know what to avoid.

PREPARATION

Preparation is very important for a successful survey, especially if the survey is done out of town. Hours can be saved because the surveyor has an accurate compass. Having the proper preparation also allows the user to do a "blind" survey. A "blind" path survey allows a surveyer to determine whether a path has line-of-sight or not without actually having to see one end from the other. This is done by walking the path, actually following the straight-line transmission path as obtained from a terrain map, identifying the heights of any potential blockages along the way. This method is exceptionally useful for surveying long paths and is well suited to surveying under nonideal weather conditions, where visibility is very limited. Only proper preparation allows the surveyor this alternative.

The most important part of the preparation is getting ready with all the surveying tools. These tools consist of

1. *Binoculars, preferably one with 7 5 35 or 8 5 50 power.* A spotting scope of similar power with zooming feature is also useful. Human eyes are effective within a 2- to 3-mile range, any distance beyond that demands a binocular resolution and magnification.
2. *Terrain contour map and street map.* These allow easy recognition and location of other building structures in the area making it easier to identify potential obstacles.
3. *Camera with telephoto lenses.* Since very few people have photographic memory, photographs always come in handy in recording the actual site information. It also allows the surveyor to refer to the pictures for answer if something was overlooked during the survey. Pictures are excellent means for preparing the contractor scope of work as well.

4. *Tape measure.* Useful to record cable distance requirement, potential antenna height, roof top dimensions, and so on.

5. *Compass, or transit.* Extremely useful for determining the path azimuth. Especially if the surveyor is not familiar with the area. The compass should be accurate to ±0.5 degree and comes with an adjustment for magnetic and true north.

6. *Mirrors (1 sq ft in size) for path flashing.* One of the more reliable ways of verifying line-of-sight for the path.

7. *Portable radios for two-way communications.*

8. *Calculator.*

Aside from surveying tools, the surveyor should have the following information ready:

1. *Path profile.* This shows the terrain profile right underneath the transmission path. Any high points along this path with a potential of blocking the transmission beam are noted and should be checked during the survey.

2. *Site information, such as building addresses, location longitude and latitude, and path azimuth.* All these can be obtained from the terrain map.

3. *Appointment with a building electrician and/or manager.* They can provide information on the best route to run any cables, the easiest way to tap power, and so on. Also the surveyor can find out at the same time whether there is any special requirement and paperwork on installing the antennas .

ACTUAL FIELD SURVEY

The goal of a successful field survey is more than determining a line-of-sight for the radio system. The objective is to collect as much information as possible to facilitate the detail-planning phase of the project, and determining a line-of-sight is just one of the many activities. This section will examine the different activities that will lead to a successful field survey.

Determining Line-of-Sight with Two Surveyors

If there are two people available for the site survey, the best approach is to assign one of them to each terminal and flash the path. Flashing a path means reflecting sunlight off a large mirror to the opposite end of the path so that the reflection is observed there. To do this

1. Depending on the time of day, the mirror should be used at one of the two locations where maximum sunlight reflection can be obtained.
2. The transmission path azimuth from this location is read off the map.
3. Knowing the path azimuth, locate the path visually by means of a compass. Note that the compasss must be adjusted to compensate for magnetic north since any value read off a map represents true north, not magnetic north.
4. Mark the path direction by placing an object a few feet in front of the surveyor, right along the azimuth.
5. Use the mirror to reflect the sunlight onto the reference marker, and then gradually move the reflected beam vertically upward until the beam is aiming toward the horizon.
6. If there is a line-of-sight, the second surveyor should be able to see the reflection readily.

The best time to do path flashing is during midmorning, on a sunny and clear day. The surveyor should locate the part of the roof which is best for installing the antenna and attempt to establish a flash. If this is not successful, look for other high points on the roof and repeat the flash. From far away, the flash will be very obvious and covers a large area. Make sure the source is clearing any obstacles along the path for at least 45 ft, about the height of three stories of a commercial building. Once a flashed line-of-sight has been established, raise and lower the mirrors to determine how much potential blockage is involved, that is, how quickly will the light beam disappear if the mirrors are lowered . If the flash is not successful, look for an intermediate location that can be the obstacle, checking the maps and path profile to see if there is any chance of bypassing it, either through using it as a repeater site or installing poles at both ends to raise the antenna heights.

Determining Line-of-Sight with One Surveyor

If only one surveyor is available, it is important to get a close visual impression of both buildings first. Remember although it is very easy to identify unique tall buildings, such as the Prudential Building in Boston, it is very difficult to identify a low-profile building, especially from a few miles away.

Upon familiarizing with the two end points, the rest of the survey basically has to do with trying to locate each structure at the respective ends. It is also helpful to sketch out on a piece of paper any unique neighboring buildings and their relative positions to the two end points.

Figure 6.1 shows a roof sketch example where a surveyor is trying to establish a path between two locations: X and Y. Standing at location X, the surveyor has used a compass to locate the most probable path direction. The beam is shooting right between two easily recognized structures, the Disneyland Park in Anaheim, California, and the Angel stadium. On one part of the roof, identified as location A, there seems to be a clear path reaching the receiver end, Y, as no obstacle is identified. On another part of the roof, viewing location B, a close-by shopping mall is blocking part of the horizon, although the possible receiver site can still be seen from there. Both sketches show the relative locations of the landmarks to the two viewing locations on the roof .

Upon completing the survey at location X, the surveyor returns to the building at the other end (location Y). Use the compass and the precalculated azimuth to locate the transmission direction. With an accurate compass, the result is usually dead on center, leading the surveyor directly to location X. By this time, the surveyor should compare his sketch prepared earlier to confirm the result, noting that all the sketch information reflects a 180-degree turn in direction because the surveyor is on the opposite end of the path. He'll also notice from far away that all buildings are almost alike. A precaution about using the compass: adjust the magnetic north reading of the compass to a true north reading of the reference point that the maps and azimuth calculations are based on. The surveyor should now see the path going between the Angel stadium and Disneyland again, except in this case, the Angel stadium is on the right side of the path and Disneyland is on the left. The far-end receiver, in this case location X,

FIELD SURVEY SKETCHES

SITE NAME: Location X
ADDRESS: 234 S. Hope St., Anaheim, Ca.   **DATE:** 1/23/88

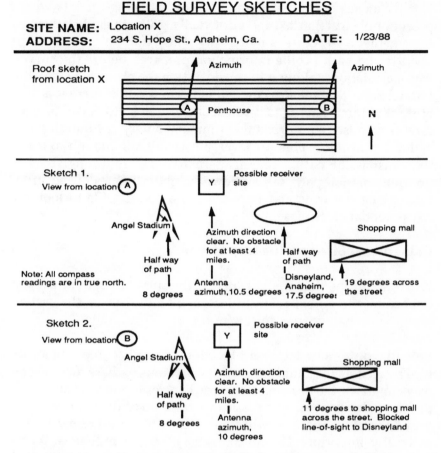

Figure 6.1. A surveyor's sketch of what he sees on the roof of location X, looking toward location Y.

should be right behind a taller building to its left, which is the shopping mall.

Determining Line-of-Sight by Walking the Path

Unfortunately, the weather does not always cooperate with the surveying activities. On a rainy, overcast day, visibility is always rather poor. In cities like Los Angeles, smog is so dense during the summer months that visibility is only good for 3 to 4 miles, and it will remain like that for weeks. The alternative in that case is to do a "blind" survey, that is, walk the path.

By means of the path profile, identify the high points where potential obstruction exists and actually go to those locations to see whether a structure is in the way. This information is then applied to the path profile information prepared before the survey trip. Calculations taking into consideration the obstacle height, the Fresnel zone, and the K factor can indicate whether there is a line-of-sight without having to confirm one visually. However, be very careful because in an urban environment, a very tall building can be located at a low contour level, but is still tall enough to block the transmission path. The best alternative is to go through the entire transmission path, taking down height estimates of the structures along the way, and impose on the path profile to look for any potential blockages.

Gathering of Site-Specific Information

While doing the line-of-sight survey, the surveyor should use the opportunity to gather relevant information to firm up the system design. We shall go over the major ones:

Potential antenna attachment location.   Based on the roof structure as well as the building owner's preference, there are various ways to attach the antenna structure. Figures 6.2 and 6.3 show the different alternatives. The most important objective is to design a sound structure that can withstand high wind conditions. This is especially important for 4-ft-diameter or larger antennas. Aside from structural issues, a wall mount is always preferred over tripod mount because roof penetration is avoided. Very often, a commercial building has a special roof warranty that requires a roof contractor to prepare and reseal the roof penetration points, which translates to additional cost and coordination effort.

When deciding the attachment location for an antenna, make sure there is ample room for the antenna to swing left and right around the azimuth to facilitate the antenna alignment process. Since one of the most popular equipment designs has the radio mounted right at the back of the antenna, make sure there is ample room behind the antenna for maintenance purposes. Look for any potential obstacle blocking the antenna. For example, it should be mounted high enough so that a 6-ft-tall person will not block the transmission if he stands right in front. In some in-

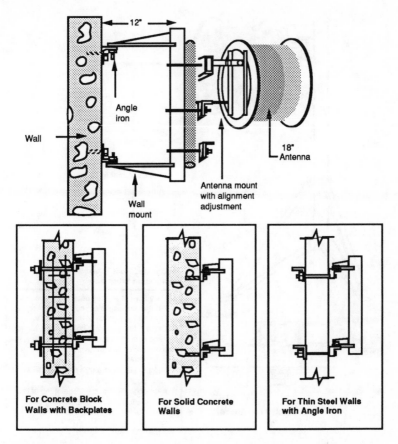

Figure 6.2.   Antenna wall mounts, showing different methods of attachments.

stances, building equipped with window washing elevators may force the antenna to be mounted as high as 10 to 15 ft above the roof level; in that case, the radio should be brought down to a suitable level, away from the antenna, for ease of maintenance (Fig. 6.4).

Aesthetic concerns. There are occasions when the building owner or the city administrators are very sensitive about the aesthetic nature of the building structure, especially one that is categorized as having historical value. When performing the site survey, it is useful to bring along a photograph of what 18-GHz and 23-GHz antennas look like, since a lot of people think of the big 10-meter satellite earth station whenever the word antenna is

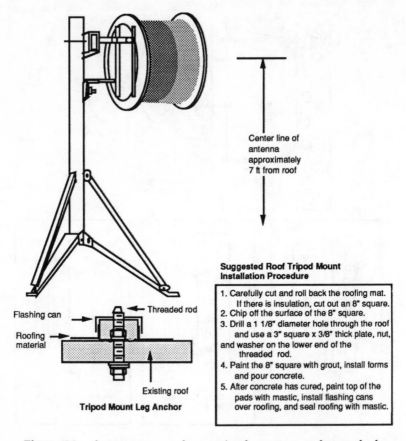

Center line of
antenna
approximately
7 ft from roof

**Suggested Roof Tripod Mount
Installation Procedure**

1. Carefully cut and roll back the roofing mat.
   If there is insulation, cut out an 8" square.
2. Chip off the surface of the 8" square.
3. Drill a 1 1/8" diameter hole through the roof
   and use a 3" square x 3/8" thick plate, nut,
   and washer on the lower end of the
   threaded rod.
4. Paint the 8" square with grout, install forms
   and pour concrete.
5. After concrete has cured, paint top of the
   pads with mastic, install flashing cans
   over roofing, and seal roofing with mastic.

Flashing can   Threaded rod

Roofing
material

Existing roof

Tripod Mount Leg Anchor

Figure 6.3. Antenna mounted on a tripod structure, to be attached to roof top.

mentioned. Unlike the lower-frequency microwave radios which require antennas of 6 to 10 foot in diameter, the common 18- and 23-GHz antennas come in the 2- to 4-ft range, which is very unobtrusive to begin with. However, if the aesthetic issue does become a problem, the possible solutions can be as simple as painting the antenna structure to blend in with the background, or even to build a small fiberglass enclosure to house the antenna (Fig. 6.5). Of course, the second alternative will mean additional cost, and a building permit is generally required.

Cable connection. All the short-haul radios we are dealing with require connection of some kind between the roof-top radio and the equipment indoors. The kinds of cables range from coaxial cable, to shielded multipair ABAM cables, to composite cable. The

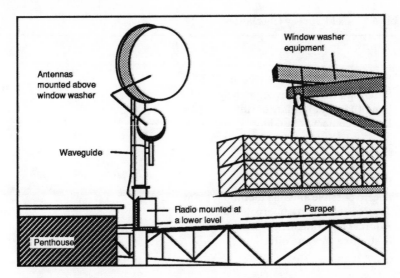

Figure 6.4. Antenna mounted high enough to clear the window washer equipment.

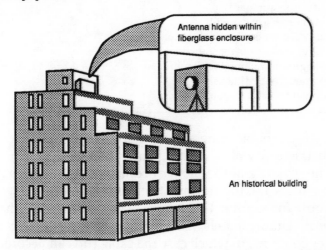

Figure 6.5. An antenna structure hidden behind a fiberglass enclosure.

dimensions are all different, depending on the various vendors. However, they all have a common requirement, which is to reach the equipment room. This means there will be installation cost on cable pull, conduit runs, floor penetrations, and so on, depending on what action is necessary. It is important to talk to someone who is familiar with the building, say, a resident electrician, and

Table 6.1   Field Survey Checklist

General:

☐ Building / support structure location
☐ Building / support structure height
☐ Longitude and latitude of site
☐ Facility contact at building

Roof top:

☐ Determine line-of-sight
☐ Locate possible antenna mounting point
☐ Type of antenna mount required
☐ Address any antenna aesthetic issues
☐ Determine antenna direction

Indoor:

☐ Estimate distance between the antenna and
     indoor equipment
☐ Locate possible cable route
☐ Identify indoor equipment location
☐ Identify possible power source
☐ Locate grounding point for equipment
☐ Equipment room dimension
☐ Equipment room location in buildin

---

identify the cabling route and the estimated cable length and find out whether any wall and floor penetration work is required.

The majority of the modern commercial buildings have telephone closets for the tenants' telephone needs. That will be the ideal route for running the radio cables and is usually the most economical. Unfortunately, some older buildings have no closet space or conduits, in which case the cable run can become very expensive. Depending on the distance between the equipment room and the roof top, it is not uncommon to run a capital expense upward of $10,000 to get the job done. From a project cost control standpoint, this requirement should be identified at the outset.

Other site surveying information.   Depending on individual users, the equipment room can range from a full-blown computer communications facility to a small closet. Take measurements of the area where the radio equipment is going to be located, including ceiling height, to make sure ample room is available. Make sure the environment, such as temperature and humidity, are within the equipment tolerance specification. Other issues are power source requirement—some radios require −48V DC source, where a separate DC power supply is necessary.

Last but not least, all the information collected during the field survey should be neatly documented. Not only is this a good engineering practice, but the documentation provides invaluable information to the vendors who are bidding on the project, and will contribute to accurate costing.

Table 6.1 provides a checklist of the different action items we have discussed so far and should be a handy tool for any future use.

CHAPTER 7
GENERAL SCOPE
OF WORK

This chapter will put together a scope of work, showing the reader what kind of information is exceptionally useful to the vendors. A turnkey vendor, like any other business, has one goal in mind: to sell a profitable system. The user, on the other hand, has an opposite objective: to put in a system at the lowest cost possible and yet satisfy all the performance objectives. It is clear that a compromise has to be made somewhere, and the best vehicle is through the use of a well-prepared request for quote (RFQ) for competitive bidding.

Very often, people have the misconception that a turnkey system RFQ is a means of passing on all the responsibility to the vendor, to make the vendor pay for any potential risks. In reality, the vendor, to cover itself from any financial exposures, will build in higher profit margin. To be competitive, the vendor will propose a very basic system that satisfies the loosely defined requirements, and nothing more. Furthermore, because the customer has perceived the project to be "risk free," the statement of work is done haphazardly, leading to misunderstanding of responsibilities, possibly project cost overrun and delays. In the end, the customer is worse off because he didn't take the time to do a thorough job in identifying what he wants.

A well-prepared request for quote allows the vendor to understand the project environment. The fewer assumptions the vendor will have to make, the more comfortable he is with the bid and the more willing to be competitive. A well-prepared request for bid should include the following technical requirements:

1. System performance objectives
2. Clearly defined vendor and client responsibilities
3. Requirements on equipment
4. Site specification identifying the project environment
5. Acceptance requirement spelling out the quality standards, pointing out what is considered satisfactory

In summary, putting in a network is very similar to purchasing a car. An intelligent buyer will do all the necessary homework to find out what features he wants in a car and then shop around for the right car at the right price. He won't walk into a dealership and buy the first car that the salesman pushes, along with all the options that are already equipped in the car.

The rest of this chapter is devoted to a generic statement of work, specifying a turnkey system. It will cover items 1 through 5 just listed. The scope of work is written to allow the vendor to provide everything, and yet it is detail enough that the vendor understands what the customer really wants. Furthermore, this scope of work is transportable: all the reader has to do is change the specifications to suit his design objective and use it for a real-life procurement.

ADVANCE TECHNOLOGY INTERNATIONAL
METROPOLITAN MICROWAVE NETWORK
SCOPE OF WORK

PROJECT OBJECTIVES

Advance Technology International (ATI) is implementing a metropolitan telecommunication network using microwave radios. This network will span four separate facilities within a 20-mile radius, carrying voice, data, and video information back and forth among each other.

The basic interface rate is 1.544 Mbps, following the North America T1 digital format. Depending on the total capacity required across each radio link, a higher data transmission speed can be used, such as T2 and T3, as long as the final interface rate is T1.

The customer premises equipment–not included in this procurement–attaching to this metropolitan network include, but are not limited to,

1. D4 channel banks with voice and 56-Kbps circuits.
2. T1 multiplexers handling synchronous and asynchronous data up to 19.2-Kbps, compressed 32-Kbps voice circuits
3. Digital PABX with T1 ports.

This network procurement requires the participating vendor to provide this on a turn-key basis, that is, to engineer, furnish, and install the necessary hardware and software to establish a microwave network to satisfy the system specification as outlined in this scope of work document.

Figure 1 shows a proposed network configuration. Also shown is the current circuit requirement of each link. Table 1 shows a more detail breakdown of circuit requirements along with a five-year growth forecast. All the participating vendors must be able to supply the equipment to support the initial circuit requirement, with provisions for growth to reach the fifth year capacity forecasted. The proposal should include the recommended system growth method, along with the corresponding upgrade cost using the current equipment and labor cost.

All the paths identified in Figure 1 have undergone a preliminary survey, and line-of-sight has been established. For comparison purposes, it is adequate to assume at this time that the network as configured in Figure 1 has:

1. Line-of-sight for all links.
2. Frequencies available for all links, meeting all carrier-to-interference objectives.
3. Met all the reliability objectives as outlined in this document.

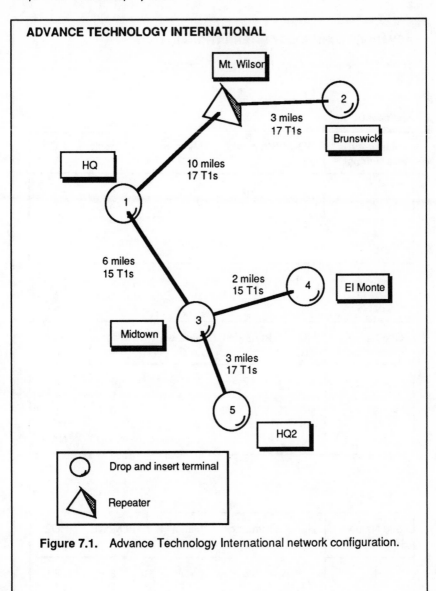

ADVANCE TECHNOLOGY INTERNATIONAL

Mt. Wilson

2

3 miles
17 T1s

Brunswick

HQ

10 miles
17 T1s

1

6 miles
15 T1s

2 miles
15 T1s

4

El Monte

3

Midtown

3 miles
17 T1s

5

HQ2

Drop and insert terminal

Repeater

Figure 7.1.   Advance Technology International network configuration.

ADVANCE TECHNOLOGY INTERNATIONAL

Table 1.   Five-Year Circuit Demand Forecast

Year 0

Site Name Site Number	HQ 1	Brunswick 2	Midtown 3	El Monte 4	HQ2 5
1		10	5	4	6
2			2	3	2
3				6	6
4					5
5					NA

Year 3

Site Name Site Number	HQ 1	Brunswick 2	Midtown 3	El Monte 4	HQ2 5
1		12	6	5	7
2			3	5	4
3				6	9
4					10
5					NA

Year 5

Site Name Site Number	HQ 1	Brunswick 2	Midtown 3	El Monte 4	HQ2 5
1		15	7	8	9
2			5	7	6
3				8	11
4					12
5					NA

NA – Not applicable.

ADVANCE TECHNOLOGY INTERNATIONAL

However, it is still the vendor's responsibility to verify these assumptions. Any adjustment in prices as a result of incorrect assumptions must be identified separately in the proposal. For example, if a vendor notices a particular path does not have a line-of-sight, the proposal should still adhere to the line-of-sight assumption, and at the same time identifies the potential problem in the proposal.

DEFINITIONS

Drop and Insert.   The capability of terminating and inserting T1 circuits.

Repeater.   The capability of repeating the RF signal through passive or active means, with or without signal regeneration.

Link.   A circuit route consists of one or more microwave paths, with drop and insert capability at both ends of the route.

Line-of-Sight.   The microwave transmission beam must clear all potential obstacles by more than 60 percent of its first Fresnel zone plus 10 ft, using a fictitious earth curvature factor (K) of 1.0.

Performance Objectives.   Performance objectives, or system objectives, are transmission performance objectives used in the engineering of network facilities.

Service Acceptance Limits.   A set of testing objectives used to determine whether the circuits are ready for service.

Network Availability.   A network may have multiple links, and all of them must satisfy the availability objective. Availability is defined as

Availability =

$$\frac{\text{Duration in a year when the T1 circuit has a BER of } 10^{-6} \text{ or better}}{\text{one year}}$$

Background Bit Error Rate.   Background bit error rate (BBER) is the ratio of the number of bits in error to the total number of bits received over a specific measurement period. BBER shows the error performance of the system under normal operation and does not include error bursts caused by line switching, maintenance activity, and so on.

Error-Free Second.   Error-free second (EFS) is the ratio of the number of seconds in which there are no error bits to the total number of seconds in the test measurement period.

Burst Error Second.   BES is 1 sec in which there are more than 100 or more error counts.

Continued

ADVANCE TECHNOLOGY INTERNATIONAL

NETWORK PERFORMANCE OBJECTIVES

The performance objectives of this microwave network are

1. Availability of 99.98 percent for each link. Outages due to multipath fading, dispersive fading, interference, rainfall, and equipment outage should be included.
2. Background bit error rate not to be higher than 10^{-8}.
3. Error-free second of 97.5 percent.
4. Jitters should meet the Bell System standards as outlined in Bell Pubs. 43802 and 43806.

The proposed system should be designed to perform at the objective level. If the vendor's system is not meeting the objectives for any reason, it should be stated in the proposal and an alternative proposed, such as using an intermediate repeater. The proposal, however, should still price out the system as specified in the statement of work, but provide an option to adopt the proposed improvement measures. A separate price, if applicable, should be included.

RESPONSIBILITY

It is the responsibility of the vendor to provide the necessary resources, including project coordinator, installation labor, engineering service, as well as the equipment and installation hardware, to establish the network.

Table 2 outlines a responsibility list. Any responsibility not covered by the list, nor discussed elsewhere in the RFQ, should be brought up by the vendor for clarification during the bidding phase. If the project is awarded, any unclear responsibility can be brought up during the engineering phase at the latest. Unforeseen obstacles encountered during the installation phase is the sole responsibility of the prime vendor. The customer is not responsible for any additional cost incurred.

SYSTEM SPECIFICATION

Circuit Interface. The interface will be at the DS1 level, with

1. Line rate of 1.544 Mbps ± 130 ppm.
2. Bipolar with at least 12.5 percent average ones density and no more than 15 consecutive zeroes.
3. Test load of 100 ohms resistance.

The circuits must conform to the DSX-1 specification as outlined in the AT&T Technical Advisory No. 34. The network must be flexible and appear transparent to valid signals appearing at its DS1 interfaces, reproducing the input signals and preserving information, format, and DS1 stream identification. It must be able to handle T1 coding schemes such as AMI, B8ZS, extended superframe, and so on.

ADVANCE TECHNOLOGY INTERNATIONAL

Table 2   SCOPE OF WORK

	RESPONSIBILITY OF	
	VENDOR	CUSTOMER
Engineering and specifications		
Real estate procurement		X
Network performance objectives		X
Network traffic requirement		X
Network configuration	X	
Establish line-of-sight and site survey	X	
Antenna support structure	X	
Cable routing plan	X	
Equipment requirement, including all necessary installation hardware, waveguide networks, etc.	X	
Contractor/subcontractor crew coordination	X	
Frequency coordination and filing permits	X	
Applying for any necessary local permits.	X	
Documentation and as built drawings	X	
Furnish equipment and installation hardware as specified	X	
Installation		
Transportation of all material to sites	X	
Antenna support structure	X	
Roof work	X	
Install and align antenna system, including any necessary waveguide plumbing	X	
All electronic equipment as specified, including powering up and test according to the acceptance criteria	X	
Conduit system	X	
Cable pull	X	
Testing	X	X

Radio

1. The equipment must be in compliance with the Federal Communications Commission (FCC) Rules and Regulations. Therefore, it must be type accepted in accordance with Rules, Part 94, as to its transmitter. Any parameters not meeting the standards but are grandfathered, or not meeting any new FCC standards to be effective in a future date, must be identified in the proposal.
2. The radio must adopt one of the widely acceptable digital transmission schemes. Analog radio using T1 transmux is not acceptable.

Continued

ADVANCE TECHNOLOGY INTERNATIONAL

3. The principal function of a protection system is to alleviate the effects of radio equipment failures. Equipment protection is provided by a hot-standby system, where the transmitter, receiver, and baseband sections are equipped with redundancy. A switch controller, also redundant, controls the switching between an active and redundant unit. The bit error rate switching threshold must be within the range of 10^{-3} and 10^{-7}.

4. The radio must be equipped with two types of alarm annunciations: LED displays showing, at minimum, warnings of power, transmitter, and receiver failures, and separate dry contact relay alarms indicating, at minimum, major and minor alarm status.

5. The office order wire equipment provides voice communications between stations. This is a required feature of the digital radio system.

6. Since the equipment is configured as hot standby, no spare parts are required. However, the vendor must provide a guaranteed 24-hr turnaround on failed parts.

Frequency Coordination.   Vendor will assist Advance Technology International in securing frequencies from the FCC. Such assistance will take the form of

1. Performing a frequency interference study and submitting a public coordination notice (PCN). Any foreign system interference must meet the radio's carrier-to-interference objectives.

2. Assisting ATI in filing for the FCC license.

3. Filing for at least four growth frequencies using the same system configuration for future expansion purpose.

Antenna and Support Structure

1. Only dual pole class A antenna should be used.

2. The antenna and its support structure must be able to sustain 85-mph (miles per hour) winds and 125-mph (miles per hour) gusts while covered by a half-inch of ice.

3. The support structure must be rigid enough so that in a high-wind situation, the antenna alignment will not be swayed so as to cause a signal strength degradation of more than 3 dB.

4. The antenna mount must come equipped with mechanical adjustment hardware to allow for fine adjustment of antenna azimuth and elevation.

5. The antenna support structure must be galvanized to prevent rusting.

6. The entire structure must be grounded to building ground.

7. The entire structure must conform to the Electronic Industries Association RS-222-C standard: Structured Standards for Steel Antenna Towers and Antenna Supporting Structures.

8. Antennas must come with protective covers or radomes. Such covers must be designed to prevent any surface icing during the winter months.

9. If a waveguide system is used to connect the antenna to the radio, and the waveguide is more than 10 ft long, a pressurized system must be used to keep the waveguide interior dry.

ADVANCE TECHNOLOGY INTERNATIONAL

10. If a waveguide system is more than 30 ft long, it must be swept tested after installation to ensure acceptable return loss.
11. The waveguide must be grounded and attached to the support structure according to the generally accepted industry standard.

Power and Grounding

1. 110 VAC/60 Hz power will be supplied at all sites. Vendor is responsible for any necessary power conversion equipment, such as DC power supply, fuse panel, and so on.
2. Battery back-up is not necessary. If this is a standard feature for the equipment, please identify the backup duration.
3. Grounding will be the typical electrical ground common to office environment. If the equipment requires a special grounding arrangement, please specify clearly.

Basic Materials and Methods

1. *Conduits.* All wiring, unless otherwise specified, shall be in conduit. Conduit inside the building shall be galvanized or sheridized rigid steel or thin wall tubing. Rigid steel conduit shall be used in poured concrete slabs and masonry walls. Conduit, installed underground and where exposed to moisture, shall be hot-dipped galvanized steel. Conduit shall be new and free from burrs and imperfections and shall be screwed together to ensure proper grounding. Any exposed conduit shall run parallel with building walls and shall be supported in a neat and substantial manner. Use junction boxes where required by code.
2. *Equipment rack.* All indoor equipment must be mounted on 7-ft bolted equipment racks supplied by the vendor. The racks are double-side, predrilled, floor-supported, steel models. Racks made by Newton Instrument Company, Inc. (Model 4010B), or others with equivalent quality is recommended.
3. *Demarcation point.* The network demarcation point at each terminal will be the vendor-supplied DSX cross-connect panel, allowing a minimum of 28 terminating positions. This demarcation point must be mounted on the same equipment rack along with the other vendor-supplied equipment. The DSX panel must have circuit monitoring features. A LED indication light is optional.

INSTALLATION GUIDELINES

General Conditions.   These specifications and the accompanying plans require the furnishing of all labor, materials, equipment, services, and permanent and temporary facilities as necessary for the communication system specified herein. This includes all work involved in the furnishing, installing, testing, adjusting, retesting, and readjusting as required and directed and placing into approved satisfactory operation all complete systems called for in these specifications and as required by job conditions.

Continued

ADVANCE TECHNOLOGY INTERNATIONAL

All labor, tools, materials, equipment, and so on, essential for the proper installation or operation of the systems which are required to provide completed systems shall be furnished by the contractor.

Coordination of All Trades. In the event that the contractor installs his work so that it interferes with the work of other trades, an ATI project manager may order the work removed and/or rearranged to eliminate the interference. Such removal and/or rearrangement, along with restoring the area previously worked on erroneously to its original condition, shall be done at the expense of the contractor.

Permits and Fees. The contractor shall give all requisite notices relative to the work when inspections are required, obtain and pay for all permits and approvals, and make all deposits for the installation of the systems as herein specified. Materials and work installed under this specification shall be in accordance with all applicable laws, codes (federal, state, and local), and ordinances.

Guarantee. The contractor shall guarantee the performance of his work for one year from the date of final acceptance. If, during this time, any defects in material or workmanship appear, they shall forthwith be corrected at no expense to ATI. This contractor shall further guarantee that all systems installed under this specification will operate in a satisfactory manner.

Rubbish. All rubbish resulting from the work herein specified shall be periodically removed by this contractor.

Safety. In accordance with generally accepted construction practices, the contractor shall be completely responsible for conditions involving his work at the job site, including safety of all persons and property during performance of his work. This requirement shall apply continuously and not be limited to normal working hours.

Insurance. The contractor must carry liability insurance in the amount of $1,000,000, covering Advance Technology International.

DOCUMENTATION

The vendor must provide documentation at least three weeks before network cutover. The documentation will address, but not be limited to, the following major items:

> System description
> Radio equipment—theory of operation
> Administrative information
> Equipment schematic diagrams
> Installation procedures
> Proposed acceptance testing
> Equipment alignment procedures
> Operating procedures
> Recommended testing procedures

ADVANCE TECHNOLOGY INTERNATIONAL

Troubleshooting information
Removal and replacement procedures
Parts list of equipment replacement parts
Recommended test equipment
Return goods procedure

PROJECT ACCEPTANCE REQUIREMENTS

The following criteria must be met before the turnkey project is deemed completed.

Circuit testing.   Test points are the T1 outputs of each radio terminal. At least one of the outputs on each terminal must undergo a testing period of 24 hr. All the network links must satisfy the minimum circuit testing standards. A network link is defined as a T1 circuit traversing over the proposed network, between any two serving terminals.

The testing parameters are:

Background BER 10^{-10} or better
Error free seconds   97.5% or better
Burst error second   1 per day maximum
Availability   99.98% or better
Testing equipment   Bit error rate test set with paper tapeoutput

Functional testing.

1. Successive simulation of equipment outages, causing corresponding protection switching.
2. All alarm indicators working.
3. Order wire functional; set to proper VF channel signal strength level.

Visual requirement.

1. All equipment must be installed neatly, with no dangling wiring.
2. All cables must be dressed and tied to the supporting structure.
3. All wire wraps and soldering work must be clean and neat. Working areas must be cleared of all debris. Vendor is responsible for clearing out all rubbish. Any openings or damages to flooring and wall must be resealed and restore to former appearance.

Continued

ADVANCE TECHNOLOGY INTERNATIONAL

SITE SPECIFICATION

General

Site name: _____   Contact name: _____
Address: _____   Title: _____
         _____   Phone no.:   (   ) _____

Longitude: _____   Ground elevation: _____
                                                          ft
Latitude: _____   Building height: _____ ft
                                                         _____ Stry.

Site characteristics

Equipment room location: _____
                                        _____
Approx. cabling distance   _____ ft
from equipment room to roof

                                        YES   NO   NOTE

Any conduits will be furnished by customer   ____   ____
Union labor required for installation   ____   ____
Building electrician required for any cable pull   ____   ____
Restrictions in working hours   ____   ____
Special restrictions
  (e.g., no wall mount for antenna)   ____   ____

Receive Site Information

Site name: _____   _____

                                        _____
Distance to receiver   _____   _____   _____
No. of T1s required   _____   _____   _____

Note: It is the responsibility of the vendor to verify all site data. Customer is not responsible for any changes.

CHAPTER 8
PROJECT COORDINATION

We have touched upon many facets of a metropolitan area network, from first conception to basic equipment design, and from network engineering to implementation. Depending on the size of the network and the experience of the information network manager, a project can take as little as a couple of months or as long as one year to complete. All the functions this book has covered thus far are important ones and should not be neglected. Furthermore, they are all dependent on each other, where each one has its contribution and priority to the overall project. This chapter will collect all the functional pieces, put together a coordinated project activities chart, and guide the project manager through the implementation phase.

To set the stage, assume that a network manager at a *Fortune 100* company has just received internal approval to establish a private metropolitan area network in Los Angeles, with the understanding that a successful undertaking will lead to similar projects in 10 other major metropolitan centers. Since this is the first private microwave network the network manager is involved with, he has to take extra precaution in running the project. Fortunately, as a starter, the Los Angeles project only involves putting in one microwave link, with each terminal located on company-leased facilities.

After reading this book, the network manager has put together a project timetable using the Critical Path Method (CPM) to keep track of the different project activities. Figure 8.1 shows the CPM chart. The project was conceptually approved on January 4, 1988, with the understanding that final budget approval is required after the capital investment figures are available. Each box represents a unique activity. Each activity, except the first one, has a predecessor. No activity can take place unless the predecessor activities have all been completed. The numbers indicated next to each activity box represents the duration in business days of each activity. According to the model, the project will be completed in 117 business days, or approximately five months.

The activities are listed in more detail in the following, along with the necessary resources involved:

ACTIVITY

1. Internal Project Approval.   Resource: None
   Issue a directive approving the feasibility study of a private microwave network project. No capital budget is approved until a more solid number is submitted. Approval on using internal resources to "fast track" the project.
2. Network Configuration Design.   Resource: Engineering
   Design the microwave network, including the setting of objectives, path design, preliminary equipment review, reliability estimates, and so on.
3. Field Survey.   Resource: Engineering
   Survey network nodes, establish line-of-sight, and review the kind of site work involved to put in a microwave system.
4. Scope of Work and RFQ Preparation.
                                    Resource: Engineering
   Assemble all the results from the design and field survey work to issue a request for quote. The scope of work should include as much information as possible (make generic assumption if possible) to maintain a consistent reply.
5. Issue RFQ.   Resource: Engineering, Vendor
   Send out the RFQ to qualified vendors to solicit bidding.

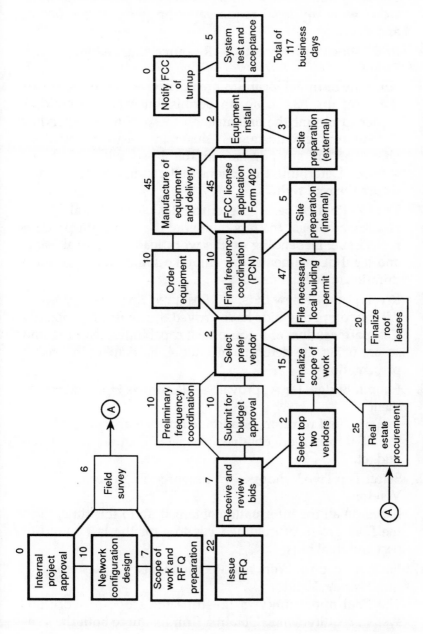

Figure 8.1. PERT chart showing activities in coordinated and logical sequence. Number adjacent to box represents activity duration.

Normally takes 22 working days, or one month to complete. The RFQ states the exact format the bids must conform to, along with any legal requirement on performance bonds, and so on.

6. **Real Estate Procurement.**   Resource: Engineering, Real Estate
Since the terminal locations are not owned by the company, the company has to obtain permission from the building owner to mount the antennas on the roof top. Depending on different circumstances, this may mean a verbal go ahead or an amendment to the lease with additional charges. Determine the actions and conditions required to secure the roof right.

7. **Finalize Roof Lease.**   Resource: Real Estate, Legal
Finalize any legal matters involving the roof-right procurement issue. Depending on the work load of the real estate and legal departments, this can take some time. Assume a duration of 20 business days.

8. **Receive and Review Bids.**   Resource: Engineering
All the competing bids have arrived. Evaluate all proposals. Compare vendors and equipment capabilities, features, and so on, using the determinant model. Participate in vendor presentations.

9. **Submit Budget for Approval.**   Resource: Upper Management
Prepare final recommendation to upper management based on all information obtained to date. Request for capital budget.

10. **Select Top Two Vendors.**   Resource: Engineering, Vendor
Based on all the information obtained up to this time, select the final group of qualified vendors for the last round of best and final bids.

11. **Finalize Scope of Work and Requirements.**   Resource: Engineering, Vendor
The final opportunity to fine-tune the scope of work and system requirements. Obtain a firm fix quote both the vendor and customer can live with. Clarify all unknowns. Work

closely with vendors on this phase to ensure clear understanding of each other's responsibilities.

12. Preliminary Frequency Coordination.   Resource: Engineering
Coordinate through a frequency coordination company the microwave link(s). Use the most preferred vendor up to that point of the vendor selection process. Determine whether there is any interference problem. Resolve any problem immediately. Reserve frequencies through the PCN process.

13. Select Prefer Vendor.   Resource: Engineering, Upper Management
Select the final vendor of choice. Obtain upper management approval.

14. Order Equipment.   Resource: Engineering, Accounting, Legal
Process purchase order on the turnkey system. Review any contracts with the legal department.

15. Final Frequency Coordination (PCN).   Resource: Engineering
Issue the final public coordination notice on the network if there is a difference with the preliminary version. Obtain the supplemental showing for the FCC filing from the frequency house as soon as possible. Specify frequencies selected to the radio vendor.

16. File Necessary Local Building Permit.   Resource: Engineering, Vendor
If required, file building permit. Prepare structural drawings for filing purposes.

17. Manufacturing of Equipment.   Resource: Vendor
The radio supplier manufactures, performs bench testing, and ships the equipment to the customer.

18. FCC License Application.   Resource: Engineering
File Form 402 with the FCC to apply for fixed radio license. Include supplemental showing prepared by the frequency house.

19. Site Preparation (Internal).   Resource: Engineering, Vendor
Complete any internal building preparation work, such as

putting in cable conduits. Any work related to the antenna structure or radio equipment installation cannot proceed without a FCC license.

20. Site Preparation (External).   Resource: Engineering, Vendor
    Building the antenna support structure, setting up the antenna, and building the fiberglass enclosure (only if necessary).

21. Equipment Installation.   Resource: Engineering, Vendor
    Assemble the equipment, power up, and antenna alignment.

22. Notify the FCC of Turnup.   Resource: Engineering
    Notify the FCC of the completion of the radio links.

23. System Test and Acceptance.   Resource: Engineering, Vendor
    Perform system testing and acceptance according to the specifications. Resolve any possible equipment failure problem. Cook the equipment for several consecutive days to clear out any infant mortality problem. Prepare the system for live traffic.

After defining the activities and their duration, the project manager can determine where the critical path is, which is the series of activities that determines the duration of the project. This is highlighted in bold lines in Figure 8.1. It shows that the time it takes from conception to vendor selection can take as long as three months and another three months to complete the system. Furthermore, the critical path takes three parallel routes during the implementation phase, which cover

1. Equipment shipment.
2. FCC license application.
3. Local building permit application.

Any delay to either one of the three parallel activities can make the critical path even longer. Using the Critical Path Method, the project manager can track closely how the project is progressing.

As mentioned earlier, this project is only the beginning. If successful, there are several other cities waiting in line for their own metropolitan networks. The project manager can learn from the experience he gains from this Los Angeles project and applies it to the other cities. Because of his new experience, the implementation time can be a lot shorter.

During the subsequent phases involving the other cities, the project engineer has decided to alter his approach to the assignment. Instead of taking all the precautionary moves, such as issuing both a preliminary and final request for proposal to detail all the requirements, he can afford to lump them into only one activity. With the experience he had gained from the previous project, he now knows exactly what the different vendors have to offer, and who are the preferred ones. Other departments in his organization have since understood what is expected in their area, such as roof-right negotiation for the real estate department and contract review for the legal people. With this in mind, the project manager puts together an updated project time schedule (Fig. 8.2). The major differences are

1. Most of the activities, especially those that are done in house, have shorter duration.

2. The preliminary RFQ and frequency coordination activities have been eliminated.

3. Certain activities can start earlier than before. For example, the first project did not begin with building permit application until the final vendor was selected, simply because the vendor had that responsibility. By moving this responsibility in house, the permit application can start earlier.

With the new experience, the original five-month project can be condensed into a four-month one. A well-run organization with an extensive experience in building metropolitan networks can even accomplish this whole project in 60 days. The key is to standardize equipment and limit the number of installation contractors as much as possible.

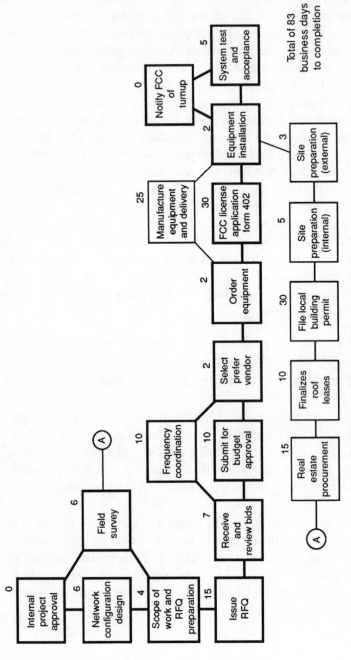

Figure 8.2. PERT chart showing simplified process.

Internal project approval — 0

Network configuration design — 6

Field survey — 6

Scope of work and RFQ preparation — 4

Issue RFQ — 15

Receive and review bids — 7

Frequency coordination — 10

Submit for budget approval — 10

Select prefer vendor — 2

Order equipment — 2

Manufacture equipment and delivery — 25

FCC license application form 402 — 30

Equipment installation — 2

Notify FCC of turnup — 0

System test and acceptance — 5

Real estate procurement — 15

Finalizes roof leases — 10

File local building permit — 30

Site preparation (internal) — 5

Site preparation (external) — 3

Total of 83 business days to completion

APPENDIX A

AN INTRODUCTION TO THE DECIBEL

The decibel, abbreviated dB, is a ratio engineers used to describe a device output and input electrical relationship. It is a logarithmic ratio expressed in the form of

$$X \text{ dB} = A \log_{10} (\text{output}/\text{input})$$

If the input and output powers of a device are compared, the formula takes the form of

$$X \text{ dB} = 10 \log_{10} (\text{power output}/\text{power input})$$

where power is measured in watts, milliwatts, and so on. If the input and output voltages are compared, the formula takes the form of

$$X \text{ dB} = 20 \log_{10} (\text{voltage output}/\text{voltage input})$$

The two formulas are essentially identical; the difference is mainly in the units in which they are expressed. The formula expressed in its first form is more commonly used in this book. For example, if a system amplifies the input signal by a hundred times, the power ratio is then

$$X \text{ dB} = 10 \log_{10} (100 / 1) = 10 \log_{10} (100) = 20 \text{ dB}$$

On the other hand, if a system attenuates the input signal by a hundred times, the power ratio then becomes

$$X \text{ dB} = 10 \log_{10} (1 / 100) = 10 \log_{10} (.01) = -20 \text{ dB}$$

It is important to know that the decibel is a ratio, not an absolute number. When we say a certain measurement is several decibels, we have to clarify the number in reference to something. For example, if we say a waveguide system loss is 5 dB, we really mean the signal power at the end of the waveguide is 5 dB lower than that at the beginning of the waveguide.

A form of the decibel unit is also used to express an absolute power-level measurement, as in the form of decibel watt and decibel milliwatt. The decibel watt is a power unit measured in reference to one watt, and decibel milliwatt is similar except it is measured in reference to milliwatts. the formulas are

$$X \text{ dBW} = 10 \log_{10} (\text{power in watts} / 1 \text{ watt})$$

$$X \text{ dBm} = 10 \log_{10} (\text{power in milliwatts} / 1 \text{ milliwatt})$$

For example, if a radio transmitter has a power output of 15 mW, the equivalent power expressed in dBm is

$$10 \log_{10} (15 \text{ mW} / 1 \text{ mW}) = 11.76 \text{ dBm}$$

The value of decibels lies in the area of ease of use. It makes calculating the network output power of a series of electrical systems very straightforward. Assume we have an electrical network that makes up of a series of black boxes, each with its characteristics of amplifications and attenuations. If these individual black boxes' amplification and attenuation factors are expressed in terms of decibels, the network's power output is simply the sum of this series of decibels. Figure A shows an example.

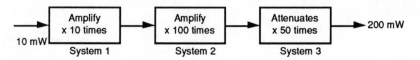

Output = 10 mW x 10 times x 100 times / 50 times = 200 mW

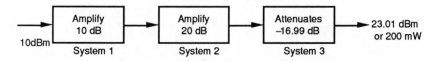

Output = 10 dBm + 10 dB + 20 dB – 16.99 dB = 13.01 dBm or 200 mW

Figure A. An input signal is fed through three system stages, experiencing amplification and attenuation. The output result is obtained through two ways of calculations.

INSTRUCTIONS FOR COMPLETING
FORM 402 APPLICATION

APPENDIX B

Approved by OMB
3060-0064
Expires 4/30/88

FEDERAL COMMUNICATIONS COMMISSION
GETTYSBURG, PA. 17325

Instructions for Completion of FCC Form 402

APPLICATION FOR STATION AUTHORIZATION IN THE PRIVATE OPERATIONAL FIXED MICROWAVE RADIO SERVICE

NOTICE TO INDIVIDUALS REQUIRED BY PRIVACY ACT OF 1974 AND THE PAPERWORK REDUCTION ACT OF 1980

Sections 301, 303, and 308 of the Communications Act of 1934, as amended, (licensing powers) authorize the FCC to request the information on this application. The purpose of the information is to determine your eligibility for a license. The information will be used by FCC staff to evaluate the application, to determine station location, to provide information for enforcement and rulemaking proceedings and to maintain a current inventory of licensees. No license can be granted unless all information requested is provided. Your response is required to obtain this authorization.

INSTRUCTIONS - General

A. CORRECT FORM:

Use FCC Form 402 to apply for an operational fixed microwave (928-929 MHz and above 952 MHz) station authorization in the Private Operational Fixed Microwave Radio Service (Part 94). This includes applications for new licenses, modifications, modifications with renewal, assignment of licenses or reinstatements of expired authorizations. For more details, see Instruction for completing Item 9A of this form.

Applications for authorizations to operate mobile microwave stations should not be made on this form. See appropriate mobile rules in the various services.

Requests for extension of time in which to construct a new station should be submitted in letter form.

B. RENEWALS:

Applications for renewal (without modification) of license, for all fixed microwave stations, should be made on FCC Form 402R, which is prepared and mailed by the FCC to licensees automatically at the appropriate time prior to the expiration date of each license.

C. FCC RULES:

Before preparing this application, applicant should refer to Part 94 of the Commission's Rules. This volume may be obtained from the Superintendent of Documents, Government Printing Office. See attached order form.

D. FEES: NO FEES ARE REQUIRED AT THIS TIME.

E. MAILING APPLICATION:

Submit completed form, properly signed and dated, to the Federal Communications Commission, Gettysburg, Pa. 17325.

INSTRUCTIONS FOR COMPLETING EACH ITEM OF THE FORM 402
SECTION I-IDENTIFICATION INFORMATION

ITEM 1. Enter the full legal name of the applicant. If you are an individual doing business in your own name, enter your name as you usually write it. If you are an individual doing business under a firm or company name (sole proprietorship), enter both your name and the firm or company name. "Doing business as" may be abbreviated as d/b/a.

Example: John Henry Doe
          d/b/a Circle Construction Company

If you are a partnership doing business under a firm or company name, enter the full name of each partner having an interest in the business, and the firm or company name.

Example: John Henry Doe and Richard Robert Roe
          d/b/a Circle Construction Company

If the application is to be submitted under the name of an unincorporated association, enter the name of the association. If the application is to be submitted under the name of a corporation or governmental entity, enter the full legal name.

ITEM 2. Enter your United States mailing address. Indicate Post Office Box and Rural Route numbers where appropriate. Be sure to include the ZIP Code. Since the address will be used to return the authorization, include the name of any person, office, or division of your organization to whom it should be addressed.

If the mailing address associated with this station has changed since your last application for this station, indicate the change by checking the box.

ITEM 3. If this application refers to a microwave station already licensed by the Commission (even if the authorization has expired) or if this is an application to relocate a presently licensed microwave station or to change, delete, or add a path at a presently licensed microwave station, enter the call sign of the station.

FCC 402 Instructions
August 1985

182

ITEM 4. If you have been assigned a Licensee Identification Number by the FCC on the last authorization for this station, enter the nine digit number. If your last authorization does not have a Licensee I.D. number or this is your first time applying for a license, leave this Item blank.

ITEM 5A. Enter the name of the person who is knowledgeable and may be contacted for clarification of the technical and administrative details of this application.

ITEM 5B. Enter this person's telephone number. Include the area code and extension.

ITEM 6. Indicate the type of applicant filing this application. Refer to the instruction for Item 1 for more details.

ITEM 7. Enter the appropriate code corresponding to the class of station (see Rule Section 94.3 for definitions) as follows:

CLASS OF STATION	CODE
Operational Fixed	FXO
Control	FX1
Fixed Relay	FX2
Operational Fixed at temporary locations	FX5

ITEM 8. Enter the rule section number from Part 81, 87, or 90 that establishes the applicant's eligibility to hold a radio license. For example, if you are applying in the Railroad Radio Service, the rule section number is 90.91. DO NOT use Rule Section 94.5 as the applicant's eligibility.

ITEMS 9A-9B. Indicate the purpose of this application as follows:

NEW STATION - Check this box and proceed to Item 10 if this is an application for a microwave station not presently licensed. If this is an application to relocate a presently licensed microwave station or to change, delete, or add a path at a presently licensed microwave station, do not check this box; refer to the instruction for Modification.

MODIFICATION - Check this box if this is an application to change the terms of a license, e.g., to relocate a presently licensed microwave station or to change, delete, or add a path at a presently licensed microwave station. See Rule Section 94.45 for a list of those changes which require you to file an application for modification of a license.

For Sections I and II: Complete Items, 1, 2, 3, 5, 6, 9, 18, and only those items that are to be changed. If the change relates to the antenna structure (Item 11-17D), then you must complete all Items 11-17D.

For Section III: Enter information only for those paths that you are changing, deleting or adding; do not fill in a column in Section III for a path that is not being changed at all. If you are changing a path, complete Items 20, 30, 31, and 32 and only those items that are to be changed in a column for the path in Section III and mark Item 9B appropriately; if you are changing any of Items 20, 30, 31, or 32, also enter in Item 9B the old value of each of these items to be changed. If you are deleting a path, complete only Items 20, 30, 31, and 32 in a column for the path in Section III and mark Item 9B appropriately. If you are adding a new path at your station, complete a new column for the path in Section III and mark Item 9B appropriately.

MODIFICATION WITH RENEWAL - Check this box if this is an application to change the terms of a license concurrent with renewal.

For Sections I and II: Complete Items 1, 2, 3, 5, 6, 9, 18, and only those items that are to be changed. If the change relates to the antenna structure (Item 11-17D), then you must complete all Items 11-17D.

For Section III: Enter information for all paths, not only for those paths that you are changing, deleting, or adding. If you are changing a path, complete a column for the path in Section III and mark Item 9B appropriately; if you are changing any of Items 20, 30, 31, or 32, also enter in Item 9B the old value of each of these items to be changed. If you are deleting a path, complete only Items 20, 30, 31, and 32 in a column for the path in Section III and mark Item 9B appropriately. If you are adding a new path at your station, complete a new column for the path in Section III and mark Item 9B appropriately.

ASSIGNMENT OF AUTHORIZATION - Check this box if this is an application to control the use and operation of a station presently authorized to another person. Attach the letter from the assignor of this authorization which is required by Rule Section 94.27(b). Complete all Items 1 through 10. If any other items are to be changed, also refer to the instruction and check the box for Modification.

OTHER - Check this box and enter the reason in the space provided if this is an application for some other reason, e.g., Reinstatement, Amendment of an application on file at the FCC, etc. For a Reinstatement, fill out the application the same as for a New Station. If this is an Amendment:

For Sections I and II: Complete Items 1, 2, 3, 5, 6, 9, 18, and only those items that are to be changed. If the change relates to the antenna structure (Items 11-17D), then you must complete all Items 11-17D.

For Section III: Enter information only for those paths that you are changing, deleting or adding; do not fill in a column in Section III for a path that is not being changed at all. If you are changing a path, complete Items 20, 30, 31, and 32 and only those items that are to be changed in a column for the path in Section III and mark Item 9B appropriately; if you are changing any of Items 20, 30, 31, or 32, also enter in Item 9B the old value of each of these items to be changed. If you are deleting a path, complete only Items 20, 30, 31, and 32 in a column for the path in Section III and mark Item 9B appropriately. If you are adding a new path at your station, complete a new column for the path in Section III and mark Item 9B appropriately.

ITEM 9C. Describe all changes being made to the station which were not covered in Item 9B. You may reference your cover letter.

ITEM 10. Indicate whether this station is to be shared by another person or persons. If it is, refer to Section 94.17 of the Commission's Rules and submit the required information with this application if it is not already on file with the FCC.

SECTION II-ANTENNA INFORMATION

ITEMS 11A-11E. Enter the requested information about the transmitting antenna location. If the location of the antenna does not have a street address, describe the location in such a way that it can be located readily. For example, if the station is on a mountain, give the name of the mountain; for antennas at rural locations, indicate the route numbers of the nearest highway intersection and the distance and direction from the nearest town.

Example: 1.3 mi N.N.W. of Erie, Pa.

Enter the names of the county and state in which the transmitting antenna structure is actually located. However, enter the name of the city that is closest to the structure even if the city is not in the same county and/or state as the structure.

Enter the geographical coordinates in degrees, minutes and seconds, rounded to the nearest second, for the antenna location. The latitude and longitude should be accurate to plus or minus one second for the antenna location. These coordinates are an important part of the location description. Do not estimate what they might be. Consult a qualified surveyor, if necessary.

ITEMS 12A-12E. Indicate in Item 12A whether the antenna(s) for this station will be mounted on an existing antenna tower (or pole). If not, do not complete Items 12B through 12E; if so, then complete Items 12B through 12E. Enter the name of a licensee using the same tower you propose to use, the name of the radio service in which he is licensed, and his station's Call Sign.

ITEM 13. If your antenna will be mounted on a tower (or pole) which is mounted on the ground, enter the overall height (feet) above ground of this tower (or pole), including as part of this tower any antennas, lights, lightning rods, etc. If you answer Item 13, DO NOT answer Item 14.

If your antenna or antenna tower (or pole), is mounted on a supporting structure such as a building, water tower, smoke stack, etc., answer Items 14A, B and C and DO NOT answer Item 13.

ITEM 14A. Answer this Item if your antenna or antenna tower (or pole) will be mounted on a supporting structure such as a building, water tower, smoke stack, etc. Enter the overall height (feet) of only the supporting structure. Include in this height anything which is a part of the supporting structure such as an elevator shaft or penthouse for a building. Lightning rods should also be included in this height.

ITEM 14B. Enter the height (feet) that the antenna or antenna tower (or pole) (including all antennas, dishes, lightning rods, lights, etc.) will extend above the height of the supporting structure in Item 14A. If your antenna or antenna tower (or pole) will not extend above the height of the supporting structure in Item 14A, enter zero (0).

ITEM 14C. Enter the overall height (feet) above ground of the supporting structure plus the antenna or antenna tower (or pole).

Item 14A + Item 14B = Item 14C

ITEM 15. Enter the ground elevation above mean sea level (feet) at the transmitting antenna site.

ITEMS 16A-16B. Enter in Item 16A the name of the nearest airport or aircraft landing area. Enter in Item 16B the direction (N.W., S.S.E., etc.) and the distance in miles from the

antenna site to the nearest runway of the nearest airport or aircraft landing area.

ITEMS 17A-17D. Indicate whether notice of construction has been filed with the Federal Aviation Administration. Refer to Part 17 of the Commission's Rules for requirements and procedures for notification. If the FAA has been notified on Form 7460-1, enter the name of the organization it was filed under in Item 17B, the city of the FAA Regional Office where it was filed in Item 17C and the date when it was filed in Item 17D.

ITEM 18. Each application for a radio station must be examined for environmental impact. In order to facilitate this determination, each applicant must state whether a Commission grant of the proposed communication facility would be a "major action" as defined by Section 1.1305 of the Commission's Rules. Briefly, a Commission grant of an application would be a "major action" if any of the following is proposed: a) a new antenna or antenna structure whose height (including all appurtenances and lighting, if any) exceeds 300 feet; b) an increase in the height of an existing antenna or antenna structure (including all appurtenances and lighting, if any) by more than ten percent (10%) that results in a final height exceeding 300 feet; c) facilities which are to be located in an officially designated wilderness area, wildlife preserve area or a nationally recognized scenic and recreational area, or facilities which will affect sites significant in American History (Section 1.1305 (a)(6)); d) construction which involves extensive changes in surface features. Further details may be found in Section 1.1305 of the Commission's Rules.

NOTE: The facilities in the above paragraph would not be major actions if located in areas devoted to heavy industry or to agriculture which are operated in support of the industrial or agricultural enterprises with which they are associated (such as for a heavy industrial plant in a heavy industrial area, or for a farmer on his own farm). If you are claiming this exemption, attach a brief explanation in support of your position.

If the answer to Item 18 is yes, submit the required Environmental Impact Narrative Statement (EINS) along with the application. Briefly, the environmental information to be contained in the EINS is: a) a description of the facilities (including height and special design features, access roads and power lines), a description of the site, the surrounding area and its uses, and a discussion of environmental and other considerations which led to its selection; b) a statement as to the zoning classification of the site (if any) and concerning communication with or proceedings before zoning, planning, environmental, or other local, State or Federal authorities on matters relating to environmental effect; c) a statement as to whether construction of the facilities has been a source of controversy on environmental grounds in the local community; d) a discussion of the nature and extent of any unavoidable adverse environmental effects perceived by the applicant and (where adverse effects are present) a discussion of any efforts made to minimize such effects and of any alternative routes, sites or facilities which have been or might reasonably be considered.

The information submitted in the EINS should be factual (not argumentative or conclusory) and should be sufficiently comprehensive and detailed to convey an understanding of the environmental consequences and to serve as a basis for a judgement concerning their significance. Further details may be found in Section 1.1311 of the Commission's Rules.

ITEM 19. If this is an application for an existing station, enter the year the station was first licensed. If this is an application for a new station, enter the year when station operation will begin.

SECTION III — TECHNICAL INFORMATION

Read the instructions for Items 9A-9B and these instructions carefully before filling out Items 20-46. Each column, A through D, pertains to a single transmission path, emanating out from this station, whose frequency is entered in Item 20 of that column. It is suggested that you enter the frequencies first and work down.

We have provided columns for only four transmission paths on the FCC Form 402. If you have additional transmission paths emanating out from this station, use additional FCC Forms 402. Fill in only Section III on these additional forms, re-labelling the columns as E, F, G, H, I, and so on as needed, and submit them attached to the first form.

ITEM 20. Enter the frequency (MHz) of the transmitter for this transmission path. Start a new column for each different frequency or transmission path. If two or more frequencies are to be used on the same path, or if the same frequency is to be used on two or more paths, start a new column for each.

ITEM 21. Enter the full emission designator of the transmitter, composed of its necessary bandwidth in kHz and its emission type, e.g., 25F9, 800F9, 10000A9Y, 5750A5, 25000A9, etc. See Rule Sections 2.201 and 2.202 for more detailed information.

ITEM 22. Enter the type of message service to be applied to the baseband of the transmitter: enter ANA for analog, DIG for digital, VID for video, or HYB for hybrid.

For digital message service, attach an exhibit describing voice and digital channel loading, data reduction or packing schemes, digital multiplex hierarchy, aggregate bit transmission rates (bps), etc.

For video message service, skip Items 23 and 24. If a non-standard bandwidth is being requested, attach an exhibit describing the video system, and additional information being carried, and why the requested bandwidth is needed.

ITEM 23. For a new path, enter the number of 4 kHz equivalent analog voice channels contemplated to be put into service on the transmitter baseband within one year from the grant of this application. For a path being changed, enter the number of such channels in use at the time of application.

ITEM 24. Enter the number of 4 kHz equivalent analog voice channels contemplated to be put into service on the transmitter baseband in the next ten years.

TRANSMITTER INFORMATION

ITEM 25. Enter the frequency tolerance (percent) of the transmitter under normal operation.

ITEM 26. Enter the gain, over an isotropic radiator, (dBi, rounded to one decimal place) of the transmitting antenna.

ITEM 27. Enter the effective isotropic radiated power (EIRP) (dBm, rounded to one decimal place) radiated off the transmitting antenna. For a periscope antenna system, this is the anticipated EIRP radiated off its reflector.

ITEM 28. Enter the beamwidth (degrees, rounded to one decimal place) of the transmitting antenna, the angular distance between the half power points of its major lobe in the horizontal plane. For a periscope antenna system, also attach an exhibit showing the dimensions of the reflector (feet) and how the antenna system complies with the requirements of Rule Section 94.75 (b). See also Rule Section 94.75 (d).

ITEM 29. Enter the height above ground (feet) to the center of the final radiating element. For a parabolic dish antenna, this is the height to the center of the dish. For a periscope antenna system, this is the height to the center of the reflector. In all cases, this height should be less than that entered in Item 13 or 14C.

ITEM 30. Enter the polarization of the transmitting antenna. Use V for vertical or H for horizontal. For other polarizations, refer to Rule Section 94.75 (c). For a periscope antenna system, this is the expected polarization of the signal radiated off the reflector.

ITEM 31. Enter the azimuth, clockwise from True North, (degrees, rounded to one decimal place) from this station to the receive site or to the first passive repeater, if any, on this transmission path.

RECEIVE SITE INFORMATION

ITEM 32. Enter the call sign of the station at the far end of this transmission path, the station which will receive the transmissions of this transmission path on the frequency entered in Item 20. For a receive-only station, enter RO. For a new station, leave this item blank. In all cases, also answer Items 33-38.

ITEM 33. Enter the gain, over an isotropic radiator, (dBi, rounded to one decimal place) of the receiving antenna.

ITEM 34. Enter the median received signal level (dBm, rounded to one decimal place) at the input to the receiver.

ITEMS 35-36. Enter the latitude and longitude (degrees, minutes, seconds, rounded to the nearest second) of the receive site.

ITEM 37. Enter the ground elevation above mean sea level (feet) of the receive site.

ITEM 38. Enter the height above ground (feet) to the center of the receiving antenna. For a parabolic dish antenna, this is the height to the center of the dish. For a periscope antenna system, this is the height to the center of its reflector. If you use space diversity, this is the height to the center of the higher antenna.

PASSIVE REPEATER NO. 1 INFORMATION

This transmission path may have one or more passive repeaters. If you do not have any passive repeaters on this path, enter NA in Item 39. If you have one passive repeater on this transmission path, answer Items 39-46.

If you have two or more passive repeaters on this transmission path, check the box provided and use additional FCC Forms 402 or separate sheets of paper. Reference these additional forms or sheets as PR 2, PR 3, and so on, answer only Items 39-46 on each, and submit them attached to the first FCC Form 402.

ITEMS 39-40. Enter the latitude and longitude (degrees, minutes, seconds, rounded to the nearest second) of this passive repeater.

ITEM 41. Enter the ground elevation above mean sea level (feet) of the site of this passive repeater.

ITEM 42. Enter the overall height above ground (feet) of the structure on which this passive repeater is to be mounted. Include any supporting structure, such as a building, water tower, smoke stack, etc., and any antenna tower (or pole), in-

cluding any lightning rods, VHF antennas, lights, beacons, etc.

ITEM 43. For a reflector passive repeater, enter the dimensions (feet × feet) of the reflector. For back-to-back parabolic dish antennas used as a passive repeater, enter the beamwidth (degrees, rounded to one decimal place) of the emanating dish, the angular distance between the half-power points of its main lobe in the horizontal plane.

ITEM 44. Enter the height above ground to the center of the reflector or back-to-back dishes.

ITEM 45. Enter the expected polarization of the signal radiated off the reflector or back-to-back dishes. Use V for vertical or H for horizontal. For other polarizations, refer to Rule Section 94.75 (c).

ITEM 46. Enter the azimuth, clockwise from True North, (degrees, rounded to one decimal place) from this passive repeater to the receive site or to the next passive repeater, if any, on this transmission path.

EXCEPT AS PROVIDED IN RULE SECTION 94.63 (e), EACH APPLICATION FOR NEW OR MODIFIED STATION AUTHORIZATION <u>MUST</u> BE ACCOMPANIED BY A CERTIFICATION OF INTERFERENCE ANALYSIS OR COORDINATION PURSUANT TO THE APPLICABLE RULE SECTIONS IN PART 94. ANY TECHNICAL PATH DATA SHEETS SUBMITTED WITH THE REQUIRED CERTIFICATION <u>MUST</u> BE CONSISTENT IN DETAIL WITH THE TECHNICAL INFORMATION PROVIDED ON THE FCC FORMS 402.

ATTACH A FUNCTIONAL SYSTEM DIAGRAM AND DETAILED DESCRIPTION OF THE MANNER IN WHICH INTER-RELATED STATIONS WILL OPERATE REQUIRED PURSUANT TO RULE SECTION 94.31 (b).

ATTACH ANY SUPPLEMENTAL EXHIBITS REQUIRED BY THE RULES OR AS REQUESTED IN THESE INSTRUCTIONS, PURSUANT TO RULE SECTION 94.31 (f).

READ THE CERTIFICATION CAREFULLY BEFORE YOU SIGN AND DATE THIS APPLICATION.

BE SURE ALL NECESSARY ATTACHMENTS ARE INCLUDED AND MAIL TO:

FEDERAL COMMUNICATIONS COMMISSION
GETTYSBURG, PA 17325

APPLICATION MUST BE SIGNED AND DATED

COMMISSION FIELD ENGINEERING OFFICES

Mailing addresses for Commission Field Engineering Offices are listed below. Street addresses can be found in local directories under "United States Government". All communications with Field Offices should be addressed to the Engineer in Charge.

Alaska, Anchorage 99510
California, Long Beach 90807
California, San Diego 92041
California, San Francisco 94111
Colorado, Denver 80228
Florida, Miami 33166
Florida, Tampa 33607
Georgia, Atlanta 30309

Hawaii, Honolulu 96850 P.O. Box 50023
Illinois, Chicago 60068
Louisiana, New Orleans 70130
Maryland, Baltimore 21201
Massachusetts, Boston 02109
Michigan, Detroit 48010
Minnesota, St. Paul 55101
Missouri, Kansas City 64133
New York, Buffalo 14202

New York, New York 10014
Oregon, Portland 97204
Pennsylvania, Philadelphia 19047
Puerto Rico, San Juan 00918-2251
Texas, Dallas 75243
Texas, Houston 77002
Virginia, Norfolk 23502
Washington, Seattle 98174

FCC 402 Instructions — Page 6
August 1985

ORDER FORM

Please send _____ copy/copies of Volume 5 of the FCC Rules containing Part 94.
S/N 004-000-00412-4- Loose leaf volume. Prices are subject to change without notice. Certain credit card charges are authorized by GPO. For information, write to the address below, or call GPO Order Desk at (202) 783-3238.

NAME–FIRST, LAST

COMPANY NAME OR ADDITIONAL ADDRESS LINE

STREET ADDRESS

CITY   STATE   ZIP CODE

PLEASE PRINT OR TYPE

☐ Remittance Enclosed (Make checks payable to Superintendent of Documents)

☐ Charge to my Deposit Account No._____

MAIL ORDER FORM TO:

Superintendent of Documents
Government Printing Office
Washington, D. C. 20402

INDEX

A

ABAM cable, 154
Absorption, of radio wave, 74
Acceptance requirements, 159
Active repeater, 163
Aesthetic concerns, 11, 153
Alarm, 28, 49, 167
  bit error rate, 50
  test-in-progress, 50
Alternate mark inversion (AMI), 24, 25, 29, 165
Amendment, FCC filing, 138
Amplitude modulation, 60
Analog network, 5
Analog radio, 165
Analog-to-Digital (A/D) converter, 15

Analog-to-digital conversion, 22
Analog transmission, noise, 14
Angle of incidence, 133
Antenna, 10, 11, 49
  alignment, 55
  alignment of, 167
  concentration of signal, 89
  envelope pattern, 128 – 29
  position of, 152 – 53
  radio output, 68
  stage, 87
  structural attachment, 152 – 53
Antenna discrimination, 127 – 32
Antenna information, FCC application, 141
Antenna structure, grounding of, 168